Hazardous Reagent Substitution
A Pharmaceutical Perspective

Green Chemistry Series

Editor-in-Chief:
James H. Clark, *Department of Chemistry, University of York, UK*

Series Editors:
George A. Kraus, *Iowa State University, USA*
Andrzej Stankiewicz, *Delft University of Technology, The Netherlands*
Peter Siedl, *Federal University of Rio de Janeiro, Brazil*

Titles in the Series:
1: The Future of Glycerol: New Uses of a Versatile Raw Material
2: Alternative Solvents for Green Chemistry
3: Eco-Friendly Synthesis of Fine Chemicals
4: Sustainable Solutions for Modern Economies
5: Chemical Reactions and Processes under Flow Conditions
6: Radical Reactions in Aqueous Media
7: Aqueous Microwave Chemistry
8: The Future of Glycerol: 2nd Edition
9: Transportation Biofuels: Novel Pathways for the Production of Ethanol, Biogas and Biodiesel
10: Alternatives to Conventional Food Processing
11: Green Trends in Insect Control
12: A Handbook of Applied Biopolymer Technology: Synthesis, Degradation and Applications
13: Challenges in Green Analytical Chemistry
14: Advanced Oil Crop Biorefineries
15: Enantioselective Homogeneous Supported Catalysis
16: Natural Polymers Volume 1: Composites
17: Natural Polymers Volume 2: Nanocomposites
18: Integrated Forest Biorefineries
19: Sustainable Preparation of Metal Nanoparticles: Methods and Applications
20: Alternative Solvents for Green Chemistry: 2nd Edition
21: Natural Product Extraction: Principles and Applications
22: Element Recovery and Sustainability
23: Green Materials for Sustainable Water Remediation and Treatment
24: The Economic Utilisation of Food Co-Products
25: Biomass for Sustainable Applications: Pollution Remediation and Energy
26: From C–H to C–C Bonds: Cross-Dehydrogenative-Coupling
27: Renewable Resources for Biorefineries
28: Transition Metal Catalysis in Aerobic Alcohol Oxidation
29: Green Materials from Plant Oils
30: Polyhydroxyalkanoates (PHAs) Based Blends, Composites and Nanocomposites

31: Ball Milling Towards Green Synthesis: Applications, Projects, Challenges
32: Porous Carbon Materials from Sustainable Precursors
33: Heterogeneous Catalysis for Today's Challenges: Synthesis, Characterization and Applications
34: Chemical Biotechnology and Bioengineering
35: Microwave-Assisted Polymerization
36: Ionic Liquids in the Biorefinery Concept: Challenges and Perspectives
37: Starch-based Blends, Composites and Nanocomposites
38: Sustainable Catalysis: With Non-endangered Metals, Part 1
39: Sustainable Catalysis: With Non-endangered Metals, Part 2
40: Sustainable Catalysis: Without Metals or Other Endangered Elements, Part 1
41: Sustainable Catalysis: Without Metals or Other Endangered Elements, Part 2
42: Green Photo-active Nanomaterials
43: Commercializing Biobased Products: Opportunities, Challenges, Benefits, and Risks
44: Biomass Sugars for Non-Fuel Applications
45: White Biotechnology for Sustainable Chemistry
46: Green and Sustainable Medicinal Chemistry: Methods, Tools and Strategies for the 21st Century Pharmaceutical Industry
47: Alternative Energy Sources for Green Chemistry
48: High Pressure Technologies in Biomass Conversion
49: Sustainable Solvents: Perspectives from Research, Business and International Policy
50: Fast Pyrolysis of Biomass: Advances in Science and Technology
51: Catalyst-free Organic Synthesis
52: Hazardous Reagent Substitution: A Pharmaceutical Perspective

How to obtain future titles on publication:
A standing order plan is available for this series. A standing order will bring delivery of each new volume immediately on publication.

For further information please contact:
Book Sales Department, Royal Society of Chemistry, Thomas Graham House, Science Park, Milton Road, Cambridge, CB4 0WF, UK
Telephone: +44 (0)1223 420066, Fax: +44 (0)1223 420247
Email: booksales@rsc.org
Visit our website at www.rsc.org/books

Hazardous Reagent Substitution
A Pharmaceutical Perspective

Edited by

Rakesh Kumar Sharma
University of Delhi, India
Email: rksharmagreenchem@hotmail.com

and

Rakeshwar Bandichhor
Dr Reddy's Laboratories Ltd, India
Email: rakeshwarb@drreddys.com

ROYAL SOCIETY OF CHEMISTRY

THE QUEEN'S AWARDS FOR ENTERPRISE: INTERNATIONAL TRADE 2013

Green Chemistry Series No. 52

Print ISBN: 978-1-78262-050-1
PDF ISBN: 978-1-78262-384-7
EPUB ISBN: 978-1-78801-376-5
ISSN: 1757-7039

A catalogue record for this book is available from the British Library

The Royal Society of Chemistry is a charity, registered in England and Wales, Number 207890, and a company incorporated in England by Royal Charter (Registered No. RC000524), registered office: Burlington House, Piccadilly, London W1J 0BA, UK, Telephone: +44 (0) 207 4378 6556.

For further information see our web site at www.rsc.org

Printed in the United Kingdom by CPI Group (UK) Ltd, Croydon, CR0 4YY, UK

Foreword

This book covers the contemporary topic "Hazardous Reagents Substitution" and features several aspects regarding reagents used for various chemical transformations. Despite the ready availability of many journals and access to various data bases and search engines, it is rather difficult for chemists to routinely and quickly glean informed inputs for selecting the right reagents for a desired synthetic transformation in terms of hazards involved and environmental footprint associated with it. This book, edited by two chemists with diverse backgrounds, one an experienced and knowledgeable chemist from industry involved in process innovation and research and the other from academia, captures some notable case studies where hazardous reagents are sought to be replaced by less hazardous ones. Valuable information about existing and upcoming reagents and catalysts for diverse transformations is also included in the book. The extent of information and trends in pharmaceutical development captured in this book makes it an essential resource for scientists working at the frontiers of research both in industry and in academia. All the chapters are written with considerable rigor and duly factor-in the context of each example and the choice of reagents employed in a particular synthesis. All the chapters provide the relevant background and include relevant references. The book can be regarded as a benchmark for the development and evolution of non-hazardous reagents and facilitate making a 'reagent switch' wherever considerations of sustainability, green chemistry and safety (non-hazardous handling) so warrant. Taken as a whole, this contribution has the potential to offer an extremely useful guide to arrive at safer reagents for synthesis of not only pharmaceutically relevant molecules but also for useful materials deployed in diverse domains.

The Editors of the book, Prof. Rakesh K. Sharma, University of Delhi and Dr Rakeshwar Bandichhor, Dr Reddy's Laboratories, Hyderabad, along with

Green Chemistry Series No. 52
Hazardous Reagent Substitution: A Pharmaceutical Perspective
Edited by Rakesh Kumar Sharma and Rakeshwar Bandichhor

Published by the Royal Society of Chemistry, www.rsc.org

others who have contributed chapters deserve to be complemented for their efforts and for drawing attention to the important issue of the avoidance of hazardous chemicals and materials. It is to be hoped that this endeavor will enthuse scientists in industry and academia to revisit and devise processes that are non-hazardous, which will go a long way in improving the public perception of chemistry.

Goverdhan Mehta, University Distinguished Professor
Kallam Anji Reddy, University of Hyderabad

Introduction

The developments in organic synthesis in recent years have been as dramatic as any that have occurred in laboratory sciences. The chemistry horizon has expanded and consolidated in such a way that one needs only mention a few terms to understand that chemical systems that did not exist roughly twenty years ago have now become a part of repertoire of synthetic organic chemistry pertaining to various research endeavours. The list of such terms would include the name reactions, organometallics, organocatalysis, oxidation, reduction, C–H activation, NHC, biocatalysis, *etc.* Surprisingly, the advances to the manipulations of these reaction techniques in a so called "non-hazardous/safer/greener" manner for the manufacturing of products of various interests have been unsuccessful in materializing a way forward for a sustainable world. The purpose of this book is to sensitize the chemistry fraternity to the idea that the usage of hazardous reagents in manufacturing of goods no longer fits in any business, environmental or technical propositions. It is rather mandatory or unavoidable to find alternates for all reported legendary yet hazardous reagents meant for or currently used for material production. These reagents, their by-products and demanding process conditions have found to pose a high degree of challenges in terms of safety and occupational exposure to the people (especially those are in the vicinity of operation) and imbalances to the ecosystem. This book features eight chapters as well as the foreword. Chapter I, entitled Introduction to Hazardous Reagent Substitution in the Pharmaceutical Industry, talks about substitution of hazardous reagents with less or non-hazardous ones as one of the most essential parts of risk management in the work place, whether it is a research and development or a manufacturing set-up. Chapter II, entitled Recyclability of Reagents, emphasizes the reduction and disposal

Green Chemistry Series No. 52
Hazardous Reagent Substitution: A Pharmaceutical Perspective
Edited by Rakesh Kumar Sharma and Rakeshwar Bandichhor

Published by the Royal Society of Chemistry, www.rsc.org

of the reagents meaningfully. Chapter III, entitled Recoverable Polymer-Supported DMAP Derivatives, elaborates on the significance of polymer-supported catalysis for chemists to recycle the reagents and reduce the overall carbon footprint of synthesis. Chapter IV, entitled Synthesis of Atorvastatin, describes the switch of a set of hazardous reagents to a less hazardous reagent system from first- to third-generation synthesis. Chapter V, entitled Synthesis of Raloxifene, talks about the evolutionary trend of reagent switching as newer generations of synthesis took precedence over time. Chapter VI, entitled Synthesis of Montelukast, encompasses the use of less hazardous reagents and scalable conditions in process chemistry routes. Chapter VII, entitled Development of a Safe, Scalable, Azide-Free Synthesis of 1-Aryl-1*H*-Tetrazoles using Diformylhydrazine, details the development of an improved and more efficient, safe process for the synthesis of tetrazoles. Finally, Chapter VIII, entitled New Directions From Academia, summarizes the scope and limitations of industry-academia collaborative research.

This book attempts to provide a perspective of finding non- or less-hazardous reagents for the hazardous reagents in practice which can be employed not just in the hitherto known synthesis but that can also be adopted for the synthesis of new entities.

Apurba Bhattacharya
Texas A&M University-Kingsville

Contents

Green Chemistry Series No. 52
Hazardous Reagent Substitution: A Pharmaceutical Perspective
Edited by Rakesh Kumar Sharma and Rakeshwar Bandichhor

Published by the Royal Society of Chemistry, www.rsc.org

CHAPTER 1

Introduction to Hazardous Reagent Substitution in the Pharmaceutical Industry

RAKESHWAR BANDICHHOR

Integrated Product Development, Innovation Plaza, Dr Reddy's Laboratories Ltd, Bachupally, Qutubullapur, R.R.Dist. 500090, Telangana, India
Email: rakeshwarb@drreddys.com

1.1 Role of Reagents in the Development of Organic Synthesis

What we have perceived over the years is that *in vitro* synthesis *per se* has a reputation of sharing similarities with *in vivo* chemical transformations (biochemical). Functional enzymes can be considered the most sophisticated green catalysts (a catalyst is different from reagent as it does not get consumed) found to be effective in cascading reactions in biological systems. However, the basic difference between synthesis and biosynthesis is that synthetic processes can be considered by and large inclusive of biosynthetic ones, whereas biosynthetic processes cannot include all possible synthetic transformations. Organic synthesis is a science that dictates the use of reactants, reagents (interchangeably used) and a set of materials towards yielding products. Interaction among all partners in the reaction, functional group susceptibility towards reagents, and their energies are the driving forces in synthetic events.

Green Chemistry Series No. 52
Hazardous Reagent Substitution: A Pharmaceutical Perspective
Edited by Rakesh Kumar Sharma and Rakeshwar Bandichhor

Published by the Royal Society of Chemistry, www.rsc.org

Since most of the reactions take place in solution, the selection of solvent(s) based on their dielectric constants and polarity is extremely important. There are some reactions where one of the reagents or reactants acts as a solvent. By definition, a reagent is a substance that is added to a reaction mixture to yield a chemical reaction.[1]

There are different types of reagents, *e.g.* inorganic acids, inorganic bases, organic acids, organic bases, epoxides, halides, azides, organometallics, carbenes, carbenoids, diazonium salts, hydrazines, phospines, ylides, silicon based reagents, oxidizing and reducing agents, *etc.* These reagents play pivotal roles in the manufacturing of goods of varied interests *e.g.* pharmaceuticals, commodity materials and materials coming from interdisciplinary industries for societal consumption.

1.1.1 Inorganic Material in the Synthesis of APIs

The use of inorganic materials, as one of the few essentials in chemical synthesis including the manufacturing of Active Pharmaceutical Ingredients (APIs), typically leads to waste generation. These are found to be primarily complex due to a variety of reasons, *e.g.* nature of the material, reaction conditions and unit operations. Chemical processes can generate acids, bases, aqueous or solvent liquors, and cyanides including metal wastes in liquid or slurry form. In organic synthesis waste solvents, either hazardous or non-hazardous, are usually recovered by distillation. Distillation is an excellent way of reusing and reducing liquid hazardous waste. In addition, the distillation left-over (solid residue) needs to be treated in such a way that there is no hazard left before it is dispensed as effluent. There are a number of strategies to achieve this, including the removal of solvents by steam stripping followed by microbiological treatment. Inorganic material in the chemical industry also includes a number of catalysts. The features of heterogeneous inorganic-material-based catalysis can be exploited by understanding the reactivity profile of such materials. Moreover, the same material can perform differently depending on overall unique structure and surfaces; therefore, it is important to measure these attributes and map the reactivity potential towards a variety of chemical transformations. It has become possible to characterize inorganic materials at the molecular level and leverage their catalytic potential. These inorganics also have the potential to offer hazardous reagent substitution to a great extent.[2]

1.1.2 Organic Material in the Synthesis of APIs

Manufacturing of APIs is an inevitable aspect of continuing health industry[3] and this involves the use of a myriad of organic entities to accomplish the material production task. Some of these organics will become integral parts of the molecule but most of them turn out to be unwanted ones contributing to a high Process Mass Intensity (PMI) or E-factor. These unwanted materials may be hazardous in nature as they may be toxic and to a great extent they may cause environmental imbalance due to an ever-increasing carbon footprint.

There are situations during manufacturing operations where organic and hazardous substance emissions should be controlled by appropriate control devices *e.g.* condensers, scrubbers, *etc.* Waste effluents from manufacturing operations contain organic and inorganic components, wash water, discharges from pumps, scrubbers and temperature controlling systems, and fleeting leaks and spills. These effluent chemicals may be of different chemical compositions, and toxic and/or genotoxic in nature. In order to minimize these hazardous unwanted materials one needs to design such a process that provides only the desired product along with the minimum possible unwanted materials. The challenges associated with this would offer opportunities to substitute hazardous chemicals/reagents with non-hazardous ones giving rise to safer by-products.

1.2 Process Mass Intensity (PMI)

PMI[4] is directly linked to the use of reactants/reagents, including water. Higher PMIs that are linked with hazardous reagents will have an exponentially high impact on cost, health and the environment. PMI is the ratio of the sum of inputs and desired product output as shown in Scheme 1.1.

As shown in Scheme 1.1, raw materials (starting material, reagents and solvents) A, B and C have been used with the quantities of 50, 20 and 5 kg respectively to give rise 5 kg D. The calculated PMI of 15 clearly reflects that the process is inefficient.

In another case, if this reaction outcome featured in Scheme 1.1 goes to a next step as an intermediate to afford product H (3 kg), after reacting with reaction partners E (40 kg), F (15 kg) and G (4 kg) as shown in Scheme 1.2, the overall PMI for product H will be calculated by omitting the value of D.

PMI is the biggest problem that any industry faces and the nature of the waste generated is another negative paradigm. There is no well-defined widely accepted mechanism in place to monitor the health impacts of chemical waste post its disposal in water streams. In fact, life cycle management of chemical waste—that may prove extremely hazardous even at ppm and ppb levels—is poorly established.

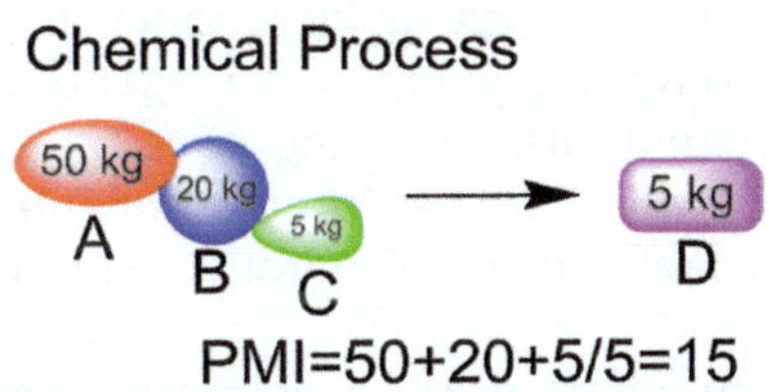

Scheme 1.1 Equation accounting for PMI towards synthesis of 'D'.

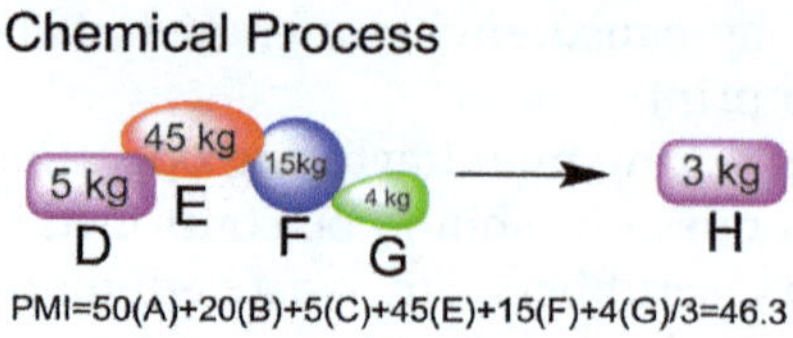

Scheme 1.2 Equation accounting for overall PMI towards synthesis of 'H'.

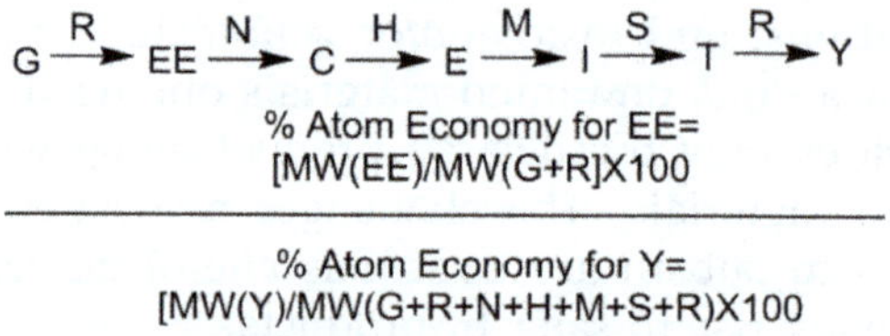

Scheme 1.3 Atom economy calculation for multistep processes.

PMI-related health hazards can never be avoided but can be minimized by chemistry and engineering excellence by design at the beginning of the process.

1.3 Stoichiometry of the Reagent

Stoichiometry and atom economy are closely associated with any chemical transformation. A highly atom economical chemical process is considered as a transformation where most of the atoms present in the reactant or reagents (but not in all cases) are incorporated in the product.[5] The atom economy is measured as a ratio of product and all reactant and reagents (when used as reactants) used multiplied by 100, and reflects that lesser amounts of reactants used is directly proportional to higher atom economy. This calculation is widely accepted for multistep processes too. Usually in such a calculation, intermediates that are formed and consumed in the next step are omitted. There are certain assumptions made about all the components of the reaction as shown in Scheme 1.3.

In this hypothetical synthesis, in order to calculate the atom economy for intermediate EE, reactants G and R are factored in, whereas the calculation of atom economy for product Y, all the reactants G, R, N, H, M, S are considered.

For instance, a reactant is considered as any material that gets incorporated into an intermediate, product or by-product during the synthesis *e.g.* certain component of protecting groups and reagents used in stoichiometric quantities (or more than that). Anything used in catalytic quantities is omitted from the calculation as they do not contribute to any of the intermediates or product(s). Solvents are also not considered as part of the atom economy calculation.

The higher the stoichiometry of the reactant/reagent, *i.e.* >1 equivalence would lead to poor atom economy and higher PMI. There are many reactions that require more than one equivalence of reactants or reagents. Historically, it was perceived that not all reagents and reactants were safe, therefore it has been imperative to design a process that would not involve stoichiometric amounts or (more than that) of these. Comparatively, catalytic processes are considered to be much safer than conventional reagent-based transformations.

1.4 Green Chemistry: Selection of Reagent

Understandably, Green Chemistry is seen as the 'right way' of doing chemistry in any phase of the process development. The business impact of Green Chemistry cannot be realised if it does not provide greener alternatives that enable a rise in optimal output in any given transformation. There are many hazardous reagents used for extremely important transformations but these are associated with high environmental impacts. One of the important areas of development in Green Chemistry is the selection of safer reagent(s), considering the nature of transformations. Reagent selection must arguably be guided by 12 Green Chemistry principles. Visiting these principles while designing the manufacturing processes could shed light on certain characteristics of reagents, allowing the selection of non-hazardous reagents. In fact, these principles do not only aid in finding safer reagent(s) but also help in reviewing the processes entirely. Chemists across both academia and industry mainly focus on achieving the highest yield in any given chemical transformation without considering anything that might add to the inefficiency of the overall process. A quest arises for the consideration of other Green Chemistry components when it is perceived that the yield is not going to be great. More often, Green Chemistry is considered last due to various reasons. In general, raw material cost, ease, and timely availability of raw material or reagent drive the decision-making in route selection. There appears to be opportunities to use the recommendations of reagent selection guidelines made available by various pharmaceutical industries in the literature. The most recent one is a very comprehensive reagent selection guideline made available by GSK.[6] During discovery research, a specific reagent is used that is not necessarily the ideal one and this provides the opportunity for chemists to use an alternative reagent for the same transformation during development followed by scale-up. One has to make smart choices while opting for alternate reagents, considering certain guidelines otherwise it amounts only to unproductive time-consuming efforts. It becomes more challenging when we deal with generics as the best possible set of reagents available in the market have already been tried by innovator companies and others; however, the newly discovered potential alternates would not have readily been available during the manufacturing of the branded medicine when there was no competition for that particular product. These new reagents need to be assessed for commercial- and manufacturing-scale viability before the entire product development and manufacturing strategies are finalized. Moreover, in order to use safer and greener alternative reagents, it is important to refer to the established reagent selection guideline toolbox and scientific rationale towards finalizing the set of reagents for any given transformation. The reagent selection guidelines are made available by considering the impact on health due to exposure of the reagents or their by-products, safety, environmental impact, their projected carbon footprint, transformational output and, last but not the least their contribution, towards over all process 'greenness'.

Biocatalysts are broadly accepted as the best reagent alternates provided such processes do not employ large amounts of water or any other solvent, demanding reaction conditions, and organic solvents in the downstream processes.

Moreover, reagent selection in my opinion is an ever-evolving science that has potential to contribute to the wellbeing of human health, business and the environment.

1.5 Positive Impacts of Hazardous Reagent Substitution During Manufacturing

The pharmaceutical industry has an increasing and lasting impact on society, in a both positive and negative sense. The negative impacts of not only pharmaceuticals but the chemical industry in general can be gauged considering the level of pollution, chemically-induced life-threating diseases, and ecological imbalance. Despite the significant amount of effort made by scientists to prevent and avoid the negative impact of chemicals on health and the environment there are certain areas that need attention. For instance, the availability of cost-effective safer or non-hazardous material, whether reactant(s), reagent(s) or product(s), is essential for business, scientists, workers, consumers and the environment.

The manufacturing of medicines is not different from any material generation at a commercial scale. However, the waste associated with pharmaceutical material production is roughly 100 kg per kg of desired product. Traditionally, by virtue of the various reaction types that are involved in the production of medicines, a number of reagents are required to effect these reactions. These reagents may not necessarily be safe to handle and the impacts of these and their by-products on health, the environment, ecosystems and food chains are not well understood. Chemists need to be inquisitive about finding alternatives for at least known hazardous reagents. Suitable safer and greener reagents for any given transformation can lead to efficient processes with lower E-factor.

The Green Chemistry tool box is considered ideal in such cases where one needs to find non-hazardous reagents for the manufacturing of medicines or materials at a large scale. Hazardous reagent substitution has the potential to contribute to sustainability as it can help minimize the generation of waste; by-products of these reagents may ultimately pose less of a risk to human health and the environment. Using non-hazardous reagents during development becomes extremely important when processes get transferred to the manufacturing facility. The scale at which generic industries operate to cater for the worldwide commercial needs of the medicine is multi-fold in comparison to the scale up batches taken during development. During process development, a number of critical parameters are identified and studied as a part of the process robustness analysis. However, a slight change in these parameters would lead to an impaired process efficiency. Due to bigger batches to obtain higher quantities of product in one go (ranging from 5 to 100 kg and sometimes more than this), a multi-fold increase in reagent quantity is unavoidable and there will be distinctive advantages if a robust process is developed by using non-hazardous reagents and solvents. These include: (1) operators may not be exposed to the hazardous chemicals; and (2) consistent output and product isolation may be simpler. In addition to this, there may not be any hazardous by-product going out as an effluent. Nevertheless, a thermodynamically stable end-product may or may not be hazardous, therefore life cycle assessment must be undertaken to understand the fate of the any chemical substance that eventually becomes a part of our ecosystem.[7]

1.6 Catalysts: Alternative Reaction Facilitators

Catalysts are considered to be alternative reaction facilitators that have functionally in common in chemical processes and in the biological system (in the case of enzymes). In general, a catalyst can be any substance that accelerates chemical transformations without being exhausted in the reaction.

Catalysts are of different types and are used in various chemical transformations. Some catalysts are derived from metals in combination with strategically-designed organic molecules (ligands), which are known as organometallic catalysts, and others are typically organic compounds that have hydrogen bonds with substrate(s) and are considered organocatalysts. Nearly all enzymes that have properties of catalysing chemical reactions are referred to as biocatalysts.

Organometallic catalysts are chemical entities that contain a metal–carbon bond and have found great application in industrial processes. Catalytic transformations can be further divided into two categories: (1) homogeneous catalysis, where all the component of the reaction including the catalysts are in same phase; and (2) heterogeneous catalysis, where the reactants and catalysts are in different phases. In most heterogeneous catalytic systems, the catalyst is in the solid phase and the reactants are liquids or gases.

The olefin metathesis reaction is one of the top-rated organometallic catalyst reactions, representing a powerful tool for the formation of carbon–carbon bonds, was discovered by Nobel Laureates, Y. Chauvin, Richard Schrock and Robert Grubbs.[8] Development of pre-catalysts by the modification of the original ruthenium complex reported by Grubbs and co-workers has led to more robust and active catalysts. These have been used in material production at a commercial scale.[9]

Other transition metal-based (mainly palladium) organometallic catalysed transformations have gained similar reputations as metathesis. Recently,

Richard F. Heck, Ei-ichi Negishi and Akira Suzuki received a Nobel Prize for advancing these catalysts for the manufacturing of materials of varied interest.[10]

Organocatalysts have also been used quite intensively and have been found to effect chemical transformation without contributing much to the waste generated after product isolation. Several research groups showed that the simple amino acid proline catalysed enantioselective cross-Aldol reactions between acetone and different aldehydes to offer corresponding products. In addition to this, chiral imidazolinium salts are found to activate α,β-unsaturated aldehydes for asymmetric Diels–Alder reactions. In general, small organic catalysts are found to effect challenging transformations. Pioneering work by MacMillan and others led organocatalysts to become one of the pillars of sustainable processes. There are other emerging organocatalysts, *e.g.* thiourea-based diversified entities, phosphates, NHCs and chiral squaramide–amine bifunctional moieties, known to be effective in the synthesis of products of varied interest including pharmaceuticals.[11]

Biocatalysts have a high degree of potential, much above to all known synthetic catalytic systems employed in the area of chemical transformations. Biocatalysis can be considered as any chemical transformation achieved by: (1) using natural or genetically engineered enzymes; (2) using fermentation technology; and (3) using a whole cell system. All these strategies have been shown to offer efficient processes with improved PMI and selectivity.[12]

Non-catalytic transformations employ stoichiometric quantities of hazardous reagents (in a significant number of cases) and exhaustive workup. These transformations tend to offer poor yields, even in some cases where chemical conversion is extremely good. This is due to repetitive muddled workup towards improving product quality by removing reagent by-product, substrate and product-related impurities. On the other hand, transformations that employ catalysts as alternatives to hazardous reagents leave behind an insignificant amount of catalyst, substrate and product-related impurities. In general, catalytic transformations provide several advantages, *e.g.* high yields, help in shortening the route, lower PMI/E-factors, higher stereo-, chemo- and regioselectivities, less demanding reaction conditions and efficient workup towards product isolation.

1.7 Nature of Reagents

There are numerous reagents known for chemical transformations. The nature of reagents can be classified based on their reactivity and hazardous features. Some of the reagents are reducing and some are oxidizing in nature. In addition to this, a few are metal based.

Not only reagents but all chemicals are classified into various hazard classes considering their physiochemical properties that are associated with negative health impacts. Each class of chemical is symbolized with a warning danger symbol as given in Table 1.1.[13]

Table 1.1 Hazardous nature of reagents.

S. No.	Reactants/reagents	Hazardous nature	Danger symbol	Remark
1	Azides, Acetylides, Diazo, Nitoso, Haloamine *etc.*	Explosive		Shock and heat sensitive and can explode
2	Hydrogen Peroxide, Hypochlorites, Nitric Acid, CrO_3, Dess–Martin periodinane *etc.*	Oxidizing		Fire or explosions occur when these come in contact with oxidizing substances
3	Ethers, $Al(O^iPr)_3$, Bu_3SnH	Flammable		At the flashpoint, the vapours on surface of liquid ignite
4	Phosgene, Hydrogen Cyanide and Nicotine, SeO_2, DDQ, $NaBH_4$	Toxic		50% of test subjects (rats, mice *etc.*) die with LD50 dose. Very toxic agents can cause damage to health at very low levels
5	Acetanilide, Ammonium Persulfates, $Mn(OAc)_3$, Glues and Resins	Harmful, Irritant, Sensitising		Cause damage to the persons but do not present high health. Cause inflammation to the skin. Respiratory and skin sensitisers
8	Acids and Caustic soda, OsO_4, NBS, LAH, Organometallic reagents	Corrosive Chemicals		If inhaled or ingested, damage can occur within the respiratory or alimentary tracts
9	Benzene, Beryllium, Asbestos, Vinyl chloride, Arsenic, Hg, Pb, Cd, Ethanol, Nitrous acid	Carcinogens, Teratogens, Mutagens		Class 1A: Known to cause cancer/non-heritable effects/ heritable genetic defects in humans 1B: Regarded as carcinogenic/ human reproductive toxins/ mutagenic to humans 2: Potential to cause carcinogenic effects/ reproductive toxin effects/ mutagenic effect
10	H_2, O_2 *etc.*	Gas in cylinder		Used to indicate gas under pressure
11	O_3, DDQ, PAHs	Toxic to Aquatic life		Dangerous to environment

Most of the reagents used for chemical transformation can be classified according to their hazardous nature. Azides, acetylides, diazo, nitoso, and haloamines are found to be explosive in nature. In addition to this, hydrogen peroxide, hypochlorites, nitric acid, CrO_3, and Dess–Martin periodinane are classified as oxidizing chemicals, whereas Ethers, $Al(O^iPr)_3$, and Bu_3SnH are considered flammable. There are some reagents, *e.g.* phosgenes, hydrogen cyanide and nicotine, SeO_2, DDQ, and $NaBH_4$, that are considered toxic. Acetanilide, ammonium persulfates, $Mn(OAc)_3$, and glues and resins are adjudged to be harmful to a great extent. Most of the organometallic reagents, along with acids and caustic soda, OsO_4, NBS, and LAH, are determined to be corrosive. Experimentally it has been proven that benzene, beryllium, asbestos, vinyl chloride, arsenic, Hg, Pb, Cd, ethanol and nitrous acid fall in the category of either carcinogens, teratogens or mutagens.

Many other hazardous reagents exist and more interestingly the question arises of what would their non-hazardous alternatives be? Recent advances in this area reveal that there are a new set of non-hazardous reagents developed for several types of chemical transformations as shown in Table 1.2.

Table 1.2 Non-hazardous reagents in different reactions.

S. No.	Type of reactions	Non/less-hazardous reagents	Remark
1	Oxidation	Manganese sulphate/Oxone/ water[14] Biocatalysts[15]	Aromatic amine to corresponding nitroso/ nitro, benzylic methyl/ alcohol to corresponding aldehyde/acid A number of biocatalysts have been identified for various oxidative transformations
2	Reduction	Reduction of different functionalities using catalyst systems[16] including biocatalysts[17]	Ketone to alcohol, nitro to amine, olefin to saturated one and reduction of difficult substrates using enzymes
3	C–C bond formation	C–C bond formation through photo generated reactive species[18] C–C bond formation at ambient temperature[19] C–C bond formation *via* C–H activation[20]	Myriad of molecules of varied interest have been synthesized
4	C–X bond formation	Transition metal catalysed C–X bond formation[21]	Functionalization of carbon centres with F, Cl, Br, I
5	C–N bond formation	C–N bond formation by using cross coupling strategies[22]	Direct inter-, intramolecular C–N bond formation to access number of molecules of varied interest

Most of the organic transformations can be categorized into five classes: oxidation; reduction; C–C; C–X; and C–N bond formation directly or through activated species. Using relatively safer reagents, the manganese sulphate/Oxone/water system is found to be a very effective oxidizing agent in converting several substrates, *i.e.* aromatic amines, benzylic methyl/alcohol, to corresponding products. A number of enzymes have also been found to effect oxidation at high yields. Reduction in general, more specifically asymmetric hydrogenations using transition-metal-based catalysts, direct reduction by using biocatalysts and chemical as well as biocatalysis-based Dynamic Kinetic Resolution (DKR), are some of the methods that have employed non-hazardous reagents in recent years. C–C, C–X and C–N bond formation are achieved by exploiting transition metals, photochemical strategies and C–H activation. The reagents used in these transformations are considered non/less-hazardous in nature.

1.8 Rationale of Chapters Recommended for the Book

Chapter 2 details the recyclability of reagents in the context of the industrial sector that has played a significant role in processing pharmaceuticals and, as a result, helped in ensuring the affordability of medicines to underprivileged societies. However, the disposal of drugs into water supplies, rivers, lakes and other waterways have put pharmaceutical chain supplies under environmental surveillance. As a consequence, a great deal of effort by the industrial researchers and engineers is now being directed towards the redesigning of existing synthetic protocols with the aid of Green Chemistry principles and practices such as elimination of hazardous substances, reuse and recycling of reagents, use of biosynthetic routes and solvent substitution practice. In this direction there have been extraordinary efforts made by several reputed companies like Pfizer, GSK, Novartis *etc.* as they have come up with some brilliant recovery and recycling strategies. However, the collective efforts of environmentalists, regulators, manufacturers and policy makers are still required to find solutions to the global environmental problem of hazardous waste management. Considering these, it is worth recommending the topic featured in Chapter 2.

Chapter 3, entitled Recoverable Polymer-Supported DMAP Derivatives, features the application of a specific polymer-supported reagent in the synthesis. The scientific fraternity worldwide is looking at E-factor reduction in synthesis as a unique opportunity to greenify the industrial processes by devising newer, safer and less/non-hazardous reagent systems. In this context, the field of polymer-supported catalysis will continue to play an important role as it allows for chemists to recycle critical species as well as reduce the overall carbon footprint of synthesis. This is especially important in the face of global climate challenges this entire generation is witnessing; in which the utilization of the principles of Green Chemistry will only become more and more vital, not only to chemistry, but to the general manufacturing practices. Although this chapter is focused on supported N-heterocyclic organocatalysts, there exist many other elegant examples underlying the importance that this field will continue to have in synthesis, and this may pave the way in finding alternative non-hazardous reagent systems for hazardous ones. This chapter is recommended considering the potentialities of polymer-supported DMAP derivatives to be recovered at a large scale if employed.

Chapter 4 discusses the synthesis of atorvastatin. Atorvastatin is marketed by Pfizer and is considered possibly the most commercially successful small molecule medicine ever launched. It is used for the treatment of hyperlipidaemia and hypercholesterolemia. The synthesis of atorvastatin evolved from the usage of hazardous reagents to the development of the processes with entirely non-hazardous reagents, therefore this chapter has been included with great interest.

Chapter 5 details the synthesis of raloxifene. Earlier synthesis of this involving hazardous reagents at a manufacturing scale posed a great challenge in terms of handling, operation and workups. These reagents were found to be unsafe and to have proven toxicity (in the majority of the cases). If not avoided, these reagents will defy the purpose of Green Chemistry in the current scenario. In this chapter, the synthesis of raloxifene is considered. Raloxifene is an estrogen agonist/antagonist, commonly considered a selective estrogen receptor modulator (SERM) centric to women health. Considering the evolving trends of switching from the use of hazardous to less-hazardous reagents as well as the use of ionic liquid in one of the syntheses as a relatively non-hazardous component, this chapter has been included.

Chapter 6 describes the synthesis of montelukast, an orally active selective leukotriene receptor antagonist that inhibits the cysteinyl leukotriene ($CysLT_1$) receptor. It was developed by Merck and Company and is currently marketed under the brand name of Singulair®. Montelukast is one of the most prescribed allergy drugs for the treatment of asthma in the United States with sales of about $4.5 billion a year (before the expiry of its patent in 2012). Merck made significant efforts to develop a commercial manufacturing process that used less-hazardous reagents. Subsequently, other generic companies also worked towards developing non-infringing and cost-effective processes, therefore this chapter has found its place in the book.

Chapter 7 details the development efforts towards a practical means to prepare 1-(4-chloro-3-fluoro-2-iodophenyl)-1*H*-tetrazole *via* reaction of an aryl diazonium salt (derived from the appropriate aniline) with diformylhydrazine. The most commonly employed methods to synthesize 1-aryl-1*H*-tetrazoles involve the use of azides, which can pose significant safety challenges. The use of diformylhydrazine affords the desired tetrazoles under conditions that are mild, convenient, safe and scalable, and circumvents the need to use azides. The other added advantage of this approach is that the entire sequence of reactions (diazotization of the amine followed by reaction with diformylhydrazine) can be carried out under aqueous conditions, rendering it a very attractive non-hazardous reagent application in the synthesis of the pharmaceutically-relevant tetrazole intermediate. These features prompted us to include this chapter in the book.

Chapter 8 features an interesting discussion on new directions from academia. The industry/academia interface for effective collaboration has great importance in the current scenario. Research and development efforts towards non-hazardous reagent substitution in the organic synthesis and processes of medicines (APIs) may effectively be possible provided multiple and meaningful interfaces exist without any barrier. Some case studies along these lines have been discussed, which encouraged us to include this chapter in the book.

1.9 Conclusions and Prospects

The substitution of hazardous reagents with less or non-hazardous ones is one of the most essential parts of risk management in the work place, whether it is a research and development or manufacturing set-up. There are certain reagents such as HCN, diazo-based reagents, phosgenes *etc.* that are restricted, or their use requires additional authorisation and protections. In general, the substitution of hazardous reagents and safety during production of medicines are non-negotiable requirements. These are only possible by usage of non or less-hazardous reagents as a result of chemistry and engineering intervention along with in-built process safety. Manufacturing processes are often linked with unchartered events due to the significant amount of heat exchange. It is important to identify the source of exothermic behaviour and have control measures in place. It is sometimes found that pressure builds up in reactors due to the fast reactivity of reagents with the substrate leading to disruption. There are other factors that govern safe practices at chemical plants, requiring systematic risk analysis of potentially hazardous chemicals and processes at scale. The risk assessment is performed considering reaction and workup temperatures, exotherm triggering temperatures, amount of pressure build-up, moisture/water sensitivity towards reactants, reagents and products, air and light sensitivity, and abrupt ignition. In addition to this, solvent compatibility, pyrophoric behaviour, peroxide formation due to the solvent or any other components of the reaction, stability towards temperature, shock and friction are also the part of hazard assessment. In order to avoid exothermic, destructive incidents, runaway reactions and exposure of hazardous reagents and chemicals, continuous manufacturing based on flow technology has taken momentum to replace the batch mode manufacturing of the products. In this direction, Novartis and MIT are running a massive collaborative project successfully.[23]

Particularly, in organic synthesis, a myriad of reagent alternatives has been discovered. Less-hazardous reagents for chemical- or biocatalysis-based asymmetric hydrogenation/reduction have been found to be very effective in the manufacturing of APIs. The USA Environmental Protection

Agency started the Presidential Green Chemistry Challenge Award, to be given to industry scientists for work towards greener processes. Some of the processes feature the replacement of hazardous reagents/catalysts to less-hazardous ones, *e.g.* in 1997, BHC Company (now BASF Corporation) synthesized ibuprofen in three catalytic steps with high atom economy, winning a Presidential Green Chemistry Challenge Award.

Boot's synthesis of ibuprofen employs reagents/reactants such as aluminium trichloride, hydroxylamine, 2-chlorobutyl ester and sodium ethoxide and none of them can be classified as non-hazardous reagents/reactants. Aluminium chloride is known to be moisture-sensitive and causes irritation to the skin and respiratory system. Hydroxyl amine is heat-sensitive and sodium ethoxide is flammable, therefore an award winning green synthesis of ibuprofen featuring catalysts (replacing hazardous reagents) has been devised. In this improved synthesis, hydrogen fluoride (catalyst and solvent), Ra-Ni and Pd catalysts are used as shown in Scheme 1.4.

This synthesis has great features as 99% of input atoms (considering the recovery of the acetic acid by-product) were incorporated into the product and overall process efficiency was improved. This is only possible due to hazardous reagents' substitution and process improvement of a synthesis which has six stoichiometric chemical transformations with 40% atom economy.[24] There are other challenging transformations that have been managed using safer reagents and safer processes and have won similar awards. In 1999, Lilly Research Laboratories developed a process for the formation of an anti-epilepsy molecule using yeast that avoided Cr-related waste.[25] Similarly, in the year 2000, Roche Colorado Corporation (Corden Pharma Colorado) worked on Ganciclovir eliminating two hazardous solid wastes and more than ten hazardous chemicals.[26] After two years, in 2002, Pfizer demonstrated outstanding work on Sertraline, doubling overall yield

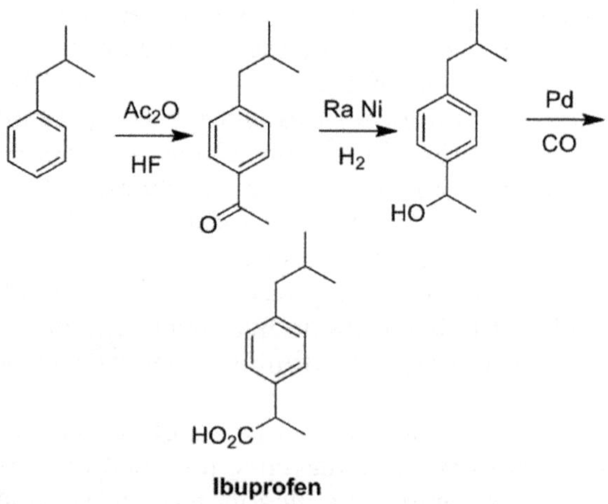

Scheme 1.4 Green synthesis of ibuprofen.

by process intensification.[27] Merck has won this award back to back in 2005 and 2006 for the highly atom economical production of Aprepitant[28] and green synthesis of β-amino acids of Sitagliptin.[29] In the same year, Codexis employed directed evolution technology towards developing biocatalysts that have been used in the synthesis of most a cost-effective chiral building block of Lipitor.[30] In 2010, Merck and Codexis jointly developed a transaminase enzyme that improved the stereoselectivity and process of the manufacture of Sitagliptin.[31] Recently, in 2014, Professor Shannon S. Stahl at the University of Wisconsin-Madison won this award through developing a Cu/aerial O_2-based catalyst system that has the potential to oxidize alcohol towards synthesizing complex APIs.[32] An ultimate attempt to make reagents safe and revolutionize industry practices is envisioned through 3D printing technology, synonymously known as additive manufacturing. This technology has the potential to carry out productions at a commercial scale. Hybridized and fabricated reaction-ware with inert polypropylene and catalyst composite-silicone material offers flexibility in the design and use for the specific needs of chemical transformation. Based on this technology, a sealed reactor can be designed to perform multistep processes which will make not only the pharmaceutical but the entire chemical industry safer and more efficient.[33] Very recently Noël's group has devised an artificial leaf considered as a reactor prototype which is potentially believed to work sustainably in the presence of sunlight as a mini-factory to produce chemical products including medicines. This technology is inspired by nature and once it becomes commercially viable this will revolutionize the chemical industry.[34]

References

1. http://ccc.chem.pitt.edu/wipf/Web/HCH.pdf.
2. Dutch National Research School Combination Catalysis Controlled by Chemical Design (NRSC-Catalysis), Eindhoven (Netherlands) April 2009.
3. P. Ball, *Nature*, 2015, **528**, 327–329.
4. W. J. W. Watson, *Green Chem.*, 2012, **14**, 251–259.
5. (a) B. M. Trost, *Science*, 1991, **254**, 1471–1477; (b) D. J. C. Constable, A. D. Curzons and V. L. Cunningham, *Green Chem.*, 2002, **4**, 521–527.
6. J. P. Adams, C. M. Alder, I. Andrews, A. M. Bullion, M. Campbell-Crawford, M. G. Darcy, J. D. Hayler, R. K. Henderson, C. A. Oare, I. Pendrak, A. M. Redman, L. E. Shuster, H. F. Sneddon and M. D. Walker, *Green Chem.*, 2013, **15**, 1542–1549.
7. Environmental Management – Life cycle assessment – Principles and Framework. EN 14040, European Committee for Standardisation, Brussels, Belgium 1997.
8. (a) T. K. Olszewski, M. Figlus and M. Bieniek, *Chim. Oggi*, 2014, **32**, 22–29; (b) R. H. Grubbs, S. J. Miller and G. C. Fu, *Acc. Chem. Res.*, 1995, **28**, 446–452.

9. (a) T. M. Trnka and R. H. Grubbs, *Acc. Chem. Res.*, 2001, **34**, 18–29; (b) R. H. Grubbs, *Handbook of Olefin Metathesis*, Wiley-VCH, Weinheim, 2003, vol. 1–3.
10. https://www.kva.se/globalassets/priser/nobel/2010/kemi/sciback_ke_10.pdf.
11. (a) R. M. D. Figueiredo and M. Christmann, *Eur. J. Org. Chem.*, 2007, 2575–2600; (b) Z. Dong, X. Jin, P. Wang, C. Min, J. Zhang, Z. Chen, H.-B. Zhou and D. Chune, *ARKIVOC*, 2011, **ix**, 367–380.
12. P. W. Sutton, J. P. Adams, I. Archer, D. Auriol, M. Avi, C. Branneby, A. J. Collis, B. Dumas, T. Eckrich, I. Fotheringham, D. Mangan and T. S. Moody "Biocatalysis in the fine chemical and pharmaceutical industries", in *Practical Methods for Biocatalysis and Biotransformations 2*, ed. J. Whitall and P. Sutton. John Wiley and Sons Ltd., 2012, 1–59.
13. (a) http://www.prc.cnrs.fr/IMG/pdf/cmr-criteria-clp.pdf; (b) http://risctox.istas.net/en/index.asp?idpagina=607; (c) *Prudent Practices in the Laboratory*, The National Academies Press, Washington, DC, 2011; the full text can be accessed free of charge at http://www.nap.edu/catalog.php?record_id=12654.
14. A. V. Vázquez Sánchez and J. G. A. Zárraga, *J. Mex. Chem. Soc.*, 2007, **51**, 213–216.
15. T. Matsuda, R. Yamanaka and K. Nakamura, *Tetrahedron: Asymmetry*, 2009, **20**, 513–557.
16. P.-G. Echeverria, T. Ayad, P. Phansavath and V. R. Virginie Ratovelomanana-Vidal, *Synthesis*, 2016, **48**, 2523–2539.
17. Z. Sun, R. Lonsdale, A. Ilie, G. Li, J. Zhou and M. T. Reetz, *ACS Catal.*, 2016, **6**, 1598–1605.
18. D. Ravelli, S. Protti and M. Fagnoni, *Chem. Rev.*, 2016, **116**, 9850–9913.
19. G. Brahmachari, *RSC Adv.*, 2016, **6**, 64676–64725.
20. M. Schnürch, *ARKIVOC*, 2015, **i**, 212–243.
21. D. A. Petrone, J. Ye and M. Lautens, *Chem. Rev.*, 2016, **116**, 8003–8104.
22. J. Bariwal and E. Van der Eycken, *Chem. Soc. Rev.*, 2013, **42**, 9283–9303.
23. https://novartis-mit.mit.edu/.
24. M. Martyn Poliakoff and P. Licence, *Nature*, 2007, **450**, 810–812.
25. B. A. Anderson, M. M. Hansen, A. R. Harkness, C. L. Henry, J. T. Vicenzi and M. J. Zmijewski, *J. Am. Chem. Soc.*, 1995, **117**, 12358–12359.
26. http://www.aspentech.com/publication_files/aiche2000.pdf.
27. P. J. Dunn, A. Wells, M. T. Williams, Green chemistry in the pharmaceutical industry, WILEY-VCH Verlag GmbH & Co, 2010.
28. M. M. Zhao, J. M. McNamara, G.-J. Ho, K. M. Emerson, Z. J. Song, D. M. Tschaen, K. M. J. Brands, U.-H. Dolling, E. J. J. Grabowski, P. J. Reider, I. F. Cottrell, M. S. Ashwood and B. C. Bishop, *J. Org. Chem.*, 2002, **67**, 6743–6747.
29. K. B. Hansen, Y. Hsiao, F. Zu, N. Rivera, A. Clausen, M. Kubryk, S. Krska, T. Rosner, B. Simmons, J. Balsells, N. Ikemoto, Y. Sun, F. Spindler, C. Malan, E. J. J. Grabowski and J. D. Armstrong, III, *J. Am. Chem. Soc.*, 2009, **131**, 8798–8804.

30. W. A. Greenberg, A. Varvak, S. R. Hanson, K. Wong, H. Huang, P. Chen and M. J. Burk, *PNAS*, 2014, **16**, 5788–5793.
31. C. K. Savile, J. M. Janey, E. C. Mundorff, J. C. Moore, S. Tam, W. R. Jarvis, J. C. Colbeck, A. Krebber, F. J. Fleitz, J. Brands, P. N. Devine, G. W. Huisman and G. J. Hughes, *Science*, 2010, **329**, 5989.
32. B. L. Ryland and S. S. Stahl, *Angew. Chem., Int. Ed.*, 2014, **53**, 8824–8838.
33. P. J. Kitson, S. Glatzel, W. Chen, C.-G. Lin, Y.-F. Song and L. Cronin, *Nat. Protoc.*, 2016, **11**, 920–936.
34. D. Cambié, F. Zhao, V. Hessel, M. G. Debije and T. Noël, *Angew. Chem., Int. Ed.*, 2017, **56**, 1050–1054.

CHAPTER 2

Recyclability of Reagents

SHIVANI SHARMA, SRIPARNA DUTTA AND
RAKESH KUMAR SHARMA*

Green Chemistry Network Centre, Department of Chemistry, University of Delhi, Delhi-110007, India
*Email: rksharmagreenchem@hotmail.com

2.1 Introduction

In today's era, pharmaceutical industries have emerged as one of the most dynamic sectors across the globe owing to innovations of new medicines and generic drugs that significantly improve human health.[1–3] Despite all these extraordinary achievements it is a harsh reality that fine chemicals with pharmacologically valuable properties are produced by liquid phase organic reactions which often lead to the generation of waste by-products like solvents and air pollutants.[4–9] The continuous flow of pharmaceutical wastes into the environment through discharge of treated wastewater, seepage from landfill sites, sewer lines, contaminated bottles, unwanted medicines and run-off from animal excretion has raised concerns about possible adverse effects on aquatic wildlife and humans (Figure 2.1). As a result, substantial efforts of pharmaceutical engineers and researchers are now directed towards the integration of green chemistry models during their synthetic processes, which not only reduce manufacturing costs but also drastically cut emissions of toxic substances.[10–15] In light of rising environmental cognizance, several attempts have been made to design sustainable approaches utilizing the basic tenets of green chemistry such as elimination of hazardous substances, reuse and recycling of reagents, use of biosynthetic routes and solvent substitution practice.

Green Chemistry Series No. 52
Hazardous Reagent Substitution: A Pharmaceutical Perspective
Edited by Rakesh Kumar Sharma and Rakeshwar Bandichhor

Published by the Royal Society of Chemistry, www.rsc.org

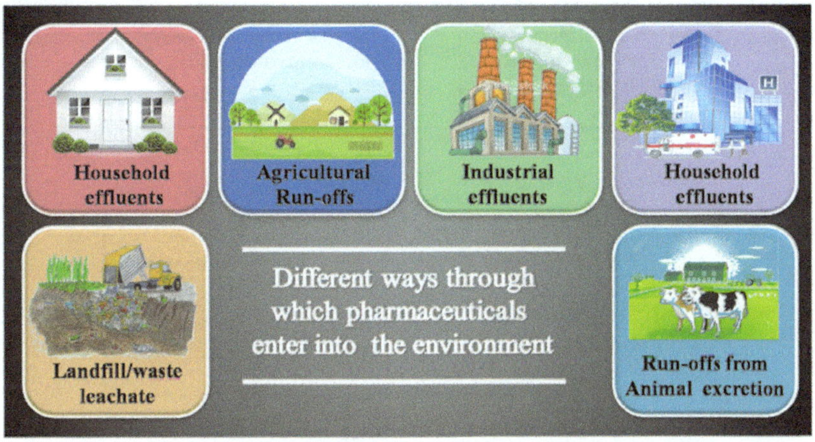

Figure 2.1 Schematic representation showing release of pharmaceutical wastes into the environment and its consequences.

Currently, recycling strategies have been widely adopted by industries and innovators worldwide as a promising solution to address waste management challenges that provide new avenue for sustainability in pharmaceutical processing sector.[16,17]

2.2 Overview of Pharmaceutical Wastes

Pharmaceutical waste may include, but is not limited to:

- Expired drugs
- Patients' personal medications
- Waste materials containing excess drugs (syringes, tubing, vials, *etc.*)
- Open drugs that cannot be used
- Containers that held drugs
- Drugs that are intended to be discarded
- Contaminated garments, absorbents and spill clean-up material, except for materials with only trace contamination

The consistent increase in the use of potent pharmaceuticals has raised serious concerns about the adverse environmental consequences due to the generation of huge amount of waste in the diagnosis, treatment of immunization and production or testing of drugs. The Environmental Protection Agency's Resource Conservation and Recovery Act (RCRA), enacted in 1976, regulates the disposal of hazardous pharmaceuticals. Depending upon the nature, character and composition, pharmaceutical wastes have been classified as hazardous, non-hazardous and mixed medical wastes (Scheme 2.1).[18–20] According to World Health Organization estimates, 85% of hospital waste is actually non-hazardous and around 10% is infectious while the remaining 5%

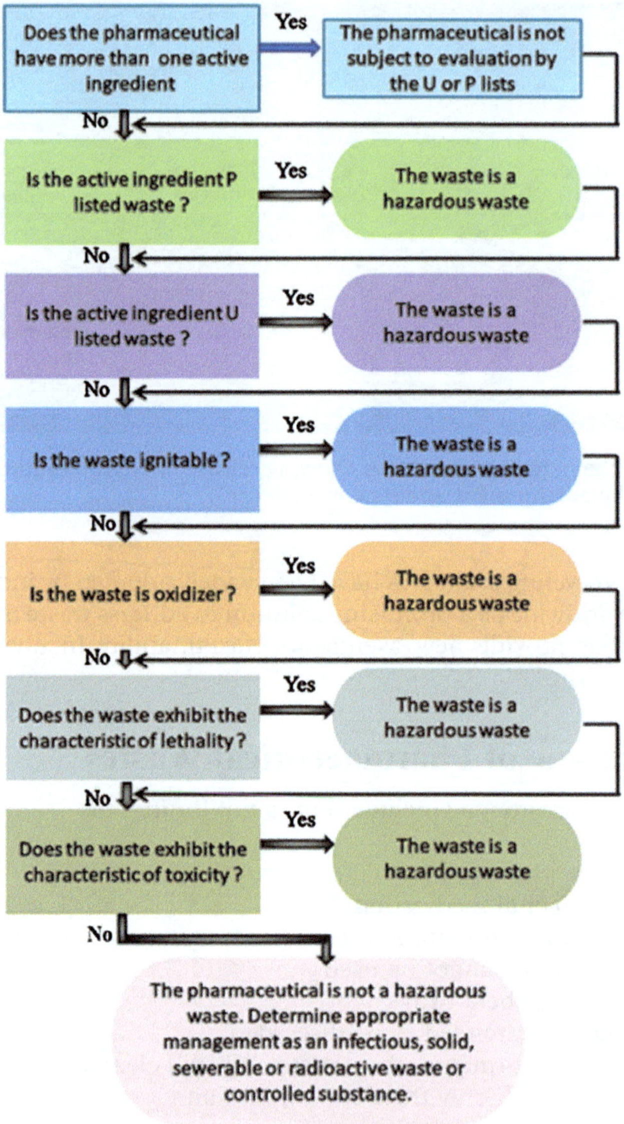

Scheme 2.1 Schematic representation of classification of pharmaceutical wastes.

is non-infectious but consists of hazardous chemicals like methyl chloride and formaldehyde.

2.2.1 Hazardous Pharmaceutical Wastes

The wastes produced by the pharmaceutical industries, in the form of a solid, liquid or gas that can pose a substantial hazard to human health or the environment when not properly managed, have been termed hazardous pharmaceutical wastes. Such types of pharmaceutical waste contain a sole active ingredient of P- or U-listed waste as described in 40 CFR Part 261, or it exhibits any hazardous properties such as ignitability, corrosivity, reactivity and toxicity.

2.2.1.1 P-listed Wastes

The P-list includes those pharmaceutical products that are categorized as "***acutely hazardous wastes***" under RCRA such as arsenic trioxide, phentermine(CIV), epinephrine, physostigmine, nicotine, physostigmine salicylate, nitroglycerin and warfarin >0.3%. Studies have revealed that an oral lethal dose of 50 mg kg^{-1} (LD50) of such P-listed wastes causes the death of 50% of a group of test animals.

2.2.1.2 U-listed Wastes

There are some drugs on the U-list. These chemicals are listed primarily for their toxicity. Similar to a P-listed waste, when a drug waste containing one of these chemicals is discarded, it must be managed as hazardous waste if two conditions are satisfied:

(1) The discarded drug waste contains a sole active ingredient that appears on the U-list, and
(2) It has not been used for its intended purpose.

2.2.1.3 Characteristics of Hazardous Wastes

Apart from the P- and U-listed categories, a waste is considered hazardous under RCRA if it possesses any of the following material characteristics:

- **Ignitability**
  In the pharmaceutical industries, a particular formulation is considered ignitable if a flammable solvent is utilized in any of the manufacturing steps. It is very important to identify whether wastes present a fire hazard while in routine storage, during disposal and transportation or if wastes are capable of exacerbating a fire once it has started. Some examples of ignitable wastes include liquids with flash points of less than 60 °C, solids that can start burning through friction or absorption of moisture, and ignitable compressed gases and oxidizers.
- **Corrosivity**
  Any pharmaceutical waste is said to be corrosive if it has a pH of less than or equal to 2 (highly acidic) or greater than or equal to 12.5 (highly basic). One of the corrosive acid wastes generated by the pharmaceutical industry is glacial acetic acid, while the examples of corrosive bases include potassium hydroxide and sodium hydroxide.

- **Reactivity**
  Pharmaceutical wastes are defined as reactive by the RCRA if they are liable to explode, or react violently to release toxic fumes when heated, compressed or mixed with water. Nitroglycerin in its pure form is the sole example that falls under this type of reactive hazardous waste.
- **Toxicity**
  A waste can be defined as "toxic" if it contains hazardous constituents that include heavy metals, pesticides, herbicides, volatile organic compounds and semi-volatile organic compounds. The ability of these hazardous constituents to leach in a landfill environment is determined by a test often termed as "Toxicity Characteristic Leaching Procedure, or TCLP." If the concentration of chemicals and heavy metals found by the TCLP exceeds the regulatory limits, the waste must be managed as hazardous waste.

2.2.2 Non-hazardous Pharmaceutical Waste

Non-hazardous pharmaceutical waste is defined as any chemical product, vaccine or allergenic which are not deemed by RCRA to be cytotoxic and cytostatic. The pharmaceutical waste which falls under this category include outdated or unused medicines that do not contain a radioactive component, discarded vials, ampules, bottles, bags and tubing containing drugs. Such types of waste may contaminate the environment through improper incineration or landfill placement, therefore proper packaging and disposal of non-hazardous pharmaceutical waste is an important challenge to be overcome.

2.2.3 How do Pharmaceuticals Enter the Environment?

There are two major pathways by which the residential and commercial pharmaceuticals can enter into the environment:

(1) **Excretion:** Excretion of drugs and metabolites by humans and animals ultimately follows sewage, septic tank, or surface run-off pathways to wastewater.
(2) **Direct Disposal:** Unused pharmaceuticals are being directly disposed to the septic tank, sewer, or landfill.

Although the pharmaceuticals which are disposed of or excreted to the sewer are sent to wastewater treatment plants, it pharmaceutical compounds cannot yet be effectively eliminated, regardless of the level of treatment (*i.e.* primary, secondary or tertiary treatment levels). On the other hand, landfill leachate can also include trace amounts of pharmaceuticals which are often sent to the same wastewater treatment systems that receive residential wastewater. Therefore, pharmaceuticals may be disposed of at engineered landfills that may merely postpone pollution of surface water and ground water. Consequently, incineration is considered to be the best method for destruction of unwanted pharmaceuticals.[21–23]

2.3 Reagents Used in Process Development

2.3.1 Solvent Utilization

Solvents play a central role in the formulation of life saving drugs as they account for roughly 56% of total mass utilized in a typical pharmaceutical batch chemical operation (Figure 2.2).[24,25]

The selection of an appropriate solvent is a major consideration in the design of active pharmaceutical ingredients (APIs) as the solvent used can enhance the yield and determine characteristics like crystal form, purity and solubility.[26] As a result, a number of solvent selection guides have been reported by GSK, Pfizer, Pharmaceutical Roundtable and by academic groups. The GSK solvent selection guide, which was developed to accommodate potential adverse effects associated with particular solvents such as inherent health or environmental hazards, has been provided in Table 2.1.

Many pharmaceutically-active products are produced *via* chemical synthesis in which solvents are often used as a carrier and in some cases, are utilized as important auxiliaries in a number of steps during manufacturing processes. These include chemical synthesis, extraction, fermentation and finishing of products. During the typical batch operation, a vessel is firstly charged with the necessary materials and after the completion of the reaction, the contents are extracted and filtered which require large quantities of solvents (Figure 2.3). Following separation, the isolated product is frequently re-crystallized in order to ensure appropriate purity and the desired crystalline form that generates solvent waste. Before any further workup, the re-crystallized product is dried which again leads to the release of toxic solvents into the environment. By far, solvents are the biggest contributors to pharmaceutical related emissions, accounting for more than 80 percent of the nearly 200 million pounds of waste emitted by the industry. Hence, several companies are focusing on the design and development of innovative strategies in order to reduce the amount of solvent used and/or the

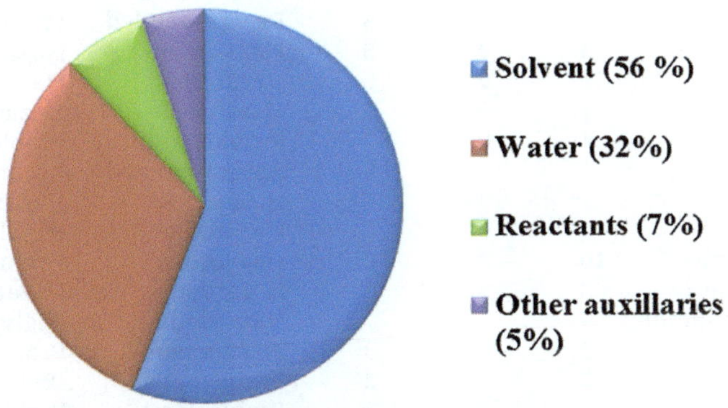

Figure 2.2 The relative contributions of materials typical in the production of an active pharmaceutical ingredient.

Table 2.1 Supporting table to highlight where issues with these solvents lie.

Solvent	Safety score	Health score	Env. score	Ranking by default	Ranking by discussion
Water	1	1	1	Recommended	Recommended
MeOH	4	7	5	Problematic	Recommended
EtOH	4	3	3	Recommended	Recommended
^{i}PrOH	4	3	3	Recommended	Recommended
^{n}BuOH	3	4	3	Recommended	Recommended
^{t}BuOH	4	3	3	Recommended	Recommended
Benzyl alcohol	1	2	7	Problematic	Problematic
Ethylene glycol	1	2	5	Recommended	Recommended
Acetone	5	3	5	Problematic	Recommended
MEK	5	3	3	Recommended	Recommended
MIBK	4	2	3	Recommended	Recommended
Cyclohexanone	3	2	5	Recommended	Problematic
Methyl acetate	5	3	5	Problematic	Problematic
^{i}PrOAc	5	2	3	Recommended	Recommended
^{n}BuOAc	4	2	3	Recommended	Recommended
Diethyl ether	10	3	7	Hazardous	Highly Hazardous
Diisopropyl ether	9	3	5	Hazardous	Hazardous
MTBE	8	3	5	Hazardous	Hazardous
THF	6	7	5	Problematic	Problematic
Me-THF	6	5	3	Problematic	Problematic
1,4 dioxane	7	6	3	Problematic	Hazardous
Anisole	4	1	5	Problematic	Recommended
DME	7	10	3	Hazardous	Hazardous
Pentane	8	3	7	Hazardous	Hazardous
Hexane	8	7	7	Hazardous	Hazardous
Heptane	6	2	7	Problematic	Problematic
Cyclohexane	6	3	7	Problematic	Problematic
Me-Cyclohexane	6	2	7	Problematic	Problematic
Benzene	6	10	3	Hazardous	Highly Hazardous
Toluene	5	6	3	Problematic	Problematic
Xylene	4	2	5	Problematic	Problematic
DCM	1	7	7	Hazardous	Hazardous
Chloroform	2	7	5	Problematic	Highly Hazardous
CCl_4	2	7	10	Hazardous	Highly Hazardous
DCE	4	10	3	Hazardous	Highly Hazardous
Chlorobenzene	3	2	7	Problematic	Problematic
Acetonitrile	4	3	3	Recommended	Problematic
DMF	3	9	5	Hazardous	Hazardous
DMAc	1	9	5	Hazardous	Hazardous
NMP	1	9	7	Hazardous	Hazardous
DMPU	1	6	7	Problematic	Problematic
DMSO	1	1	5	Recommended	Recommended
Sulfolane	1	9	7	Hazardous	Hazardous
HMPA	1	9	7	Hazardous	Highly Hazardous
Nitromethane	10	2	3	Hazardous	Highly Hazardous
Methoxyethanol	3	9	3	Hazardous	Hazardous
Carbon disulphide	9	7	7	Hazardous	Highly Hazardous
Formic acid	3	7	3	Problematic	Problematic
Acetic acid	3	7	3	Problematic	Problematic
Ac_2O	3	7	3	Problematic	Problematic
Pyridine	4	2	3	Recommended	Hazardous
TEA	6	7	3	Problematic	Hazardous

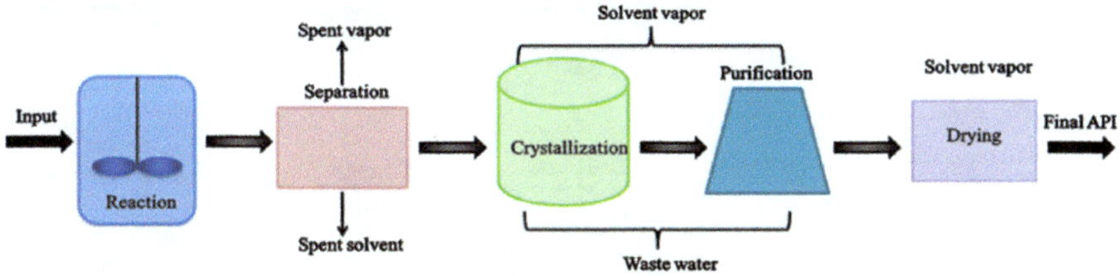

Figure 2.3 Typical pharmaceutical batch operation.

utilization of more environmentally benign solvents. Currently, two different waste solvent treatment options are generally applied: thermal treatment in hazardous waste solvent incinerators; and solvent recovery. Studies reveal that reusing solvents instead of incinerating them is a much more sustainable and greener solution that is not only important from an ecological viewpoint but also contains a real economic benefit for industrial users.

2.3.2 Role of Catalysts

Catalysis has become an increasingly important technology for the pharmaceutical industry as it enhances and expedites chemical reactions involved in the synthesis of active ingredients and intermediates during drug manufacturing processes.[27–29] These processes quite often employ precious metal catalysts such as palladium, ruthenium, rhodium, platinum *etc.* that provide a myriad of benefits over traditional reaction chemistry including better activity, higher selectivity and higher turnover numbers. However, despite these significant features, its widespread applications in practical and process chemistry are still hampered due to the associated drawbacks of high costs and difficult separation from the reaction mixture. In order to overcome these issues, several recovery techniques such as solvent extraction, nanofiltration, chemical precipitation and adsorption have been utilized by the industrial sector so far.[30–32]

2.3.2.1 Nanofiltration Technique

Organic solvent nanofiltration or solvent resistant nanofiltration represents an innovative membrane technology with enormous potential as it allows catalyst separation down to a molecular level. In this technique, solvent-stable membranes are generally utilized in order to separate larger molecules from organic solvents. During this process, when pressure is applied across a nanoporous film, the feed stream gets separated into retentate and a permeate stream as shown in Figure 2.4. So, we can say that this process is based on high permeability of the product and a low permeability of the catalyst through the membrane surface.

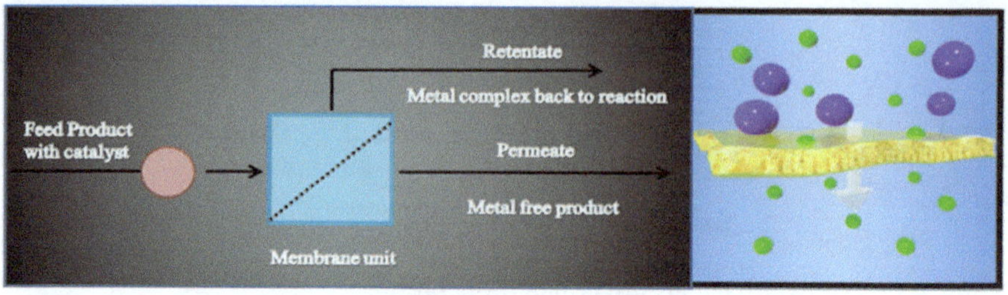

Figure 2.4 Organic solvent nanofiltration.

2.3.2.2 *Ion Exchange Resin Technology*

Over the past 60 years, ion exchange resin technology has been utilized by the biopharmaceutical industry for the separation of the catalyst from the reaction mixture. For this purpose, specialized ion exchange resins having different functional groups on their surface are frequently employed which form very strong complexes with transition metals. Subsequently, the metals can be removed from such strongly bound metal complexes using strong acidic or basic solution through a process called elution.

The resin-based catalyst removal approach has several advantages, some of which include:

- *Selective recovery of the catalyst, leading to higher recoveries of API*
- *Providing solid waste for incineration or selective desorption of the catalyst*

2.3.2.3 *Chemical Precipitation*

Chemical precipitation is a conventional technology which has been widely used to separate metal from the reaction products due to its operational simplicity. In this process, insoluble salts of metals in the form of hydroxides, sulphides, carbonates and phosphates are produced by the addition of an appropriate precipitant. For example, if we consider the hydroformylation reaction catalyzed by hydridocobalt tetracarbonyl [$HCo(CO)_4$], an insoluble salt is formed upon the addition of sodium bicarbonate which can be filtered off, while the products *i.e.* aldehyde remain in the filtrate. Subsequently, the catalyst can be reactivated through acidification of cobalt salt with hydrochloric acid.

$$HCo(CO)_4 + NaHCO_3 \rightarrow NaCo(CO)_4 + H_2CO_3$$

2.3.2.4 *Heterogenization Approach*

Amongst the enlisted methodologies the heterogenization approach, which involves the immobilization of existing homogeneous metal complexes onto solid support matrices such as polymers, inorganic support and so forth, is now being adopted by chemists and industrial researchers across

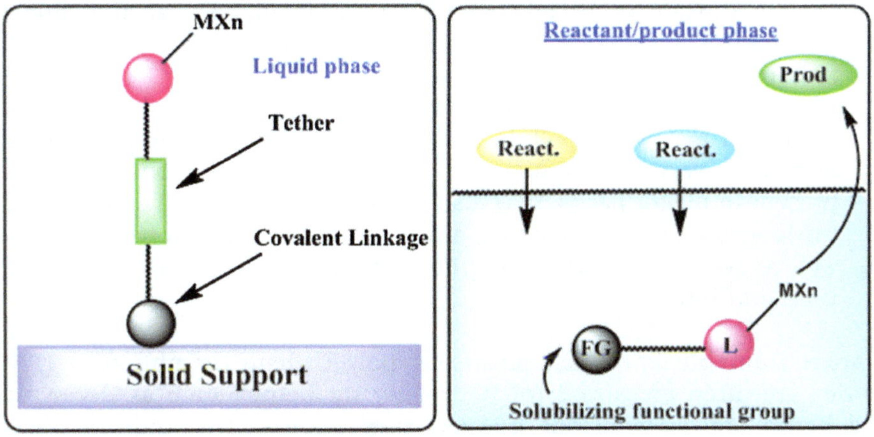

Figure 2.5 Heterogenization of homogeneous catalyst.

the globe (Figure 2.5). These solid supported heterogeneous catalysts not only reduce the loss of catalyst and enhances reusability, making the catalyst cost effective. but also display improved efficiency due to the stable active site and better steric control of the reaction intermediate that make them an interesting alternative to the conventional homogeneous and heterogeneous counterparts.

2.3.3 Pharma Packaging Materials

Pharmaceutical packaging materials are the collection of different components which not only keep the drugs free from contamination or microbial growth but also ensure their safe delivery from the industries to the hands of customer. There are several materials used for the packaging process, however 80% of the pills in the world are packed with blister packaging which consists of a lid made up of aluminium (Al) and paper and a cavity consisting of polyvinyl chloride (PVC), polypropylene (PP), polyester (PET) or aluminium (Al) (the blister). The improper disposal of these materials would have adverse effect on the environment.

- **Burning:** Burning of blister packs leads to the generation of many harmful products due to the combustion of polypropylene ($[C_3H_6]_n$), polyester ($[C_{10}H_8O_4]_n$), and polyvinyl chloride ($[CH_2CHCl]_n$) as mentioned below:

$$[C_3H_6]_n + 9n/2\ O_2 \rightarrow 3n\ CO_2 + 3n\ H_2O$$

$$[C_{10}H_8O_4]_n + 10n\ O_2 \rightarrow 10n\ CO_2 + 4n\ H_2O$$

$$[CH_2CHCl]_n + 2n\ O_2 \rightarrow nCO_2 + n\ H_2O + nHCl + n\ CO$$

Carbon dioxide and water vapour result in global warming due to greenhouse effect while HCl, which is a primary combustion product of polyvinyl chloride, causes harm to human beings.

- **Disposal:** Disposal of such packaging materials as normal waste would cause soil and water pollution as the degradation process of PVC, PP and PET is very slow. Besides, Al^{3+}, which is a key component of the lid present in blister packs, may also cause a detrimental effect on aquatic (toxic agent for gill-breathing animals like invertebrates and fish) as well as terrestrial ecosystems (disrupts metabolic processes of mammals and birds).

Apart from blister packs, plastic or polyethylene terephthalate (PET) bottles are often employed for packaging medicines such as dry syrups and liquid oral suspensions because they are easy to handle in distribution and retail processes. However, the health ministry is considering a ban on the use of such bottles for packaging pharmaceutical products due to concerns of adverse effects in the environment. Consequently, the attention of the drug manufacturing sector is now being paid towards either recycling of these materials or use of other eco-friendly packaging substitute materials such as bioplastics like plant-derived polysaccharides obtained from plant sources (corn, wheat, potatoes) and proteins which are recyclable.

2.4 Process Efficiency Metrics

The pharma industry stands out amongst manufacturing industries for the exorbitant amounts it spends every year in the design and development of highly potent drugs. However, the last decade has witnessed a steep decline in the discovery, endorsement and marketing of new chemical entities. This decline is associated with an increasing concentration of research and development investments in areas where the need is great but the risk of failure is high such as unmet therapeutic needs and unexploited biological mechanisms. To meet these unparalleled challenges, the **drug development industry** is restructuring worldwide with the goal to reduce manufacturing costs for the producers and improve operational excellence, productivity and quality. The green chemistry movement has developed a series of metrics that have been employed in the pharmaceutical sector that cover areas of resources, materials, processing, cleaning, life cycle assessment and renewability. Some of the common metrics which have been commonly utilized are defined and discussed below:[33]

2.4.1 Atom Economy

The term 'atom economy' (AE) was first coined by Barry Trost in 1991 at Stanford University with an attempt to prompt synthetic organic chemists to pursue 'green chemistry.' Simply stated atom economy is an excellent metric

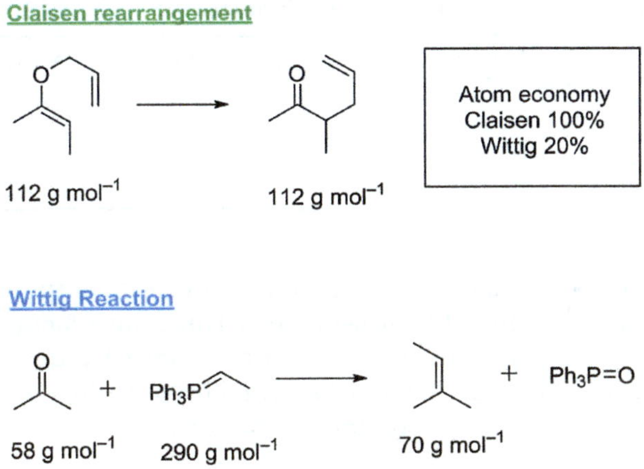

Figure 2.6 A comparison of AE of Claisen rearrangement and Wittig reaction.

to assess how efficiently a reaction has been designed with respect to the utilization of reactants. In other words, it is the calculation of how much of the reactants are converted into a desired product.

$$\text{AE} = [\text{molecular weight of product/total molecular weight of reactants}] \times 100\%$$

An example of a reaction with perfect AE (*i.e.* 100%) is the Claisen rearrangement in which all of the atoms from the starting materials are incorporated into the product and there are no side products formed (Figure 2.6). On the contrary, a prototypical Wittig reaction has a poor AE (merely 18.5%).

The AE concept is very simple as it can be calculated readily from the reaction scheme, but it is associated with a number of drawbacks. It does not take into consideration important factors such as reaction yield and stoichiometry. Besides, it does not take into account the solvents or other reagents utilized either in the reaction or during the workup procedure.

2.4.2 E factor

E factor is the simplest and most effective green metric developed by Roger Sheldon, professor emeritus at the Delft University of Technology in Netherlands. It is defined as the ratio of waste over product and hence evaluates the "greenness" of a manufacturing route by its process efficiency. If the E factor is high, more waste would be generated which eventually leads to a negative environmental impact.

The E factor ranges for different branches of chemical industries proposed by Sheldon have been provided in Table 2.2 which clearly shows that the

Table 2.2 E factors across the chemical industry.

Industry sector	Annual production (*t*)	E factor	Waste produced
Oil refining	10^6–10^8	Ca.0.1	10^5–10^7
Bulk chemicals	10^4–10^6	<1–5	10^4–5×10^6
Fine chemicals	10^2–10^4	5–50	5×10^2–5×10^5
Pharmaceuticals	10–10^3	25–100	2.5×10^2–10^5

pharmaceutical industry as a whole has a high E factor compared with other industrial chemical manufacturing segments. Thus, the E factor has played a significant role in drawing the attention of the pharma industry worldwide to problems related to waste generation, thereby providing an impetus for the development of cleaner, sustainable processes.

2.4.3 Reaction Mass Efficiency

The term reaction mass efficiency was first introduced by GSK in 2002 for measuring the greenness of a process. It is calculated as mass of product divided by the sum of the masses of reactants in a balanced chemical equation. It takes into consideration some of the important reaction parameters such as yield, AE and stoichiometry.

There are two ways to calculate RME. For a generic reaction:

$$A + B \rightarrow C$$

$$\text{Reaction mass efficiency} = \frac{\text{M.W. of product C}}{\text{M.W. of A} + (\text{M.W. of B} \times \text{molar ratio B/A})} \times \text{Yield}$$

Or more simply:

$$\text{Reaction mass efficiency} = \frac{\text{Mass of product C}}{\text{Mass of A} + \text{Mass of B}} \times 100$$

2.4.4 Process Mass Intensity (PMI)

The fourth metric "PMI" measures the mass of materials used to obtain 1 kg of the API and is defined as the total mass of materials used to produce a specified mass of product. Total mass includes all the reagents such as reactants, solvents, catalysts, additives *etc.*

$$\text{Mass Intensity} = \text{Mass of all materials used to make product (kg)/mass of product (kg)} \times 100\%$$

The concept of PMI is quite similar to that of the E factor as it can be measured with or without inclusion of water. One of the most interesting aspects associated with PMI is that it can be calculated simply by using inputs into a reaction rather than waste measurements. In an ideal process, PMI should approach one while on the other hand E factor is zero.

$$\text{E factor} = \text{PMI} - 1$$

2.4.5 Carbon efficiency

Carbon efficiency as proposed by GSK is widely used by the pharmaceutical industries since the development of carbon skeleton is essential for their work. It is defined as the percentage of carbon in the reactants that remain in the final product which can be mathematically represented as shown:

For a generic reaction: $A + B \rightarrow C$

$$\%\ \text{Carbon efficiency} = \frac{\text{Amount of carbon in product} \times 100}{\text{Total carbon present in reactants}}$$

2.5 Waste Management Practices Across the World

2.5.1 Pharmaceutical Industry Wastewater and Its Treatment

Access to quality water is one of the pressing global challenges that humanity is facing in the present century. Water plays an indispensable role in multiple stages of pharmaceutical manufacturing, which include production, material processing, extraction, cleaning washing of equipment, maintenance, cooling *etc.* Waste water generated from these industries varies in composition as well as quantity depending on the starting materials and the processes that have been employed in the manufacturing of APIs. The complete compositions of the waste water produced by these industries have been summarized in Table 2.3.[34] Different pharmaceutical agents like analgesics, anti-inflammatories, antidepressants, antihypertensives, contraceptives, antibiotics, steroids and hormones *etc.* have been identified in water samples around the globe from the ng l^{-1} to $\mu\text{g l}^{-1}$ range. Studies have shown that even such low concentrations of these pharmaceutical residues have acute and chronic effects on various organisms including humans, animals and aquatic species. According to one of the reports from the literature, chronic exposure of fathead minnow – *"a species of fish belonging to the Pimephales genus"* to 5–6 ng l^{-1} concentrations of the synthetic estrogen 17α-ethinylestradiol in a freshwater lake resulted in collapse of the fish population in that lake owing to complete reproductive failure.[35] As a result of this, the whole lake ecosystem was affected due to disruption of the food web. Similarly, low microgram concentrations of the antidepressant drug "fluoxetine" and an anti-epilepsy drug "carbamazepine" have been found to adversely affect the behaviour of the fathead minnow and Atlantic salmon fish (*Salmo salar*).[36,37]

Thus, from the viewpoint of environmental and health concerns, regular monitoring of pharmaceutical waste water entering different drinking water sources becomes highly important.

Table 2.3 Composition of pharmaceutical waste waters.[34]

Chemical processes waste waters			Fermentation process waste waters		
Parameter	Min-max value	Avg composition	Parameter	Min-max value	Avg composition
COD, $mg\,L^{-1}$	375–32 500	8854	COD, $mg\,L^{-1}$	180–12 380	4670
BOD_5, $mg\,L^{-1}$	200–6000	2344	BOD_5, $mg\,L^{-1}$	25–6000	2150
BOD_5 : COD ratio	0.1–0.6	0.32	BOD_5 : COD ratio	0.2–0.6	0.4
TOC, $mg\,L^{-1}$	860–4940	2467	TOC, $mg\,L^{-1}$	190–760	440
TKN, $mg\,L^{-1}$	165–770	383	TKN, $mg\,L^{-1}$	65.5–190	128
NH_3-N, $mg\,L^{-1}$	148–363	244	NH_3-N, $mg\,L^{-1}$	3.3–11	7
TDS, $mg\,L^{-1}$	675–9320	6.9	TDS, $mg\,L^{-1}$	1300–28 000	12 950
pH	3.9–9.2		TSS, $mg\,L^{-1}$	57–7130	1200
			Conductivity, $\mu S\,cm^{-1}$	1600–44 850	17 800
Anions					
Cl^-, $mg\,L^{-1}$	760–4200	2820		182–2800	1500
SO_4^{2-}, $mg\,L^{-1}$	890–1500	1260		160–9000	2100

2.5.1.1 *Treatment of Pharmaceutical Wastewater*

A wide range of wastewater treatment and disposal methods are employed by the pharmaceutical industry. Five of such approaches that have been utilized to treat pharmaceutical wastewaters include:

- *Recovery of individual APIs or drugs which are likely to be present in wash waters and solvents*
- *Physico-chemical treatment by coagulation-flocculation and sedimentation*
- *Aerobic/anaerobic biological treatment*
- *Advanced oxidation processes*
- *New hybrid technologies specific to the pharmaceutical industry*

(1) **Recovery processes**
A very crucial waste control methodology adopted by pharmaceutical plants is the pretreatment and recovery of various useful by-products, including solvents, acids, heavy metals, and several important APIs, present in the effluents discharged by these industries. Recovering such pharmaceutical products can not only cut down the waste disposal costs but also reduce the waste treatment operational costs significantly, thereby improving the economics of the drug manufacturing processes. In fact, the recovered waste streams can also be utilized elsewhere in other processes. For instance:
- The recycled water could be further used for cooling of towers and other processes as well, which would eventually reduce the consumption of precious fresh water

- Even the processed hot waste streams could be used for as either the boiler feed or in the heat exchangers which also decreases the water and energy costs

When the recovery processes of API are considered, nanofiltration stands out to be one of the most promising amongst several membrane separation processes such as ultrafiltration, reverse osmosis *etc.* for the removal of pharmaceuticals having molecular weights greater than 250 Da.[38,39] Shahtalebi *et al.* have reported that this technique can be used for the recovery of amoxicillin which is an antibiotic often taken for the treatment and prevention of respiratory, gastrointestinal, urinary and skin bacterial infections from the waste streams.[40] Apart from nanofiltration technique, ultrafiltration is a cross-flow separation process that has also been widely used for the recovery of useful organic compounds from the waste water from different fermentation processes. Using this process, Bezawada *et al.* were able to recover alkaline protease which is a very important material for the fermentation industry.[41]

(2) **Physico-chemical treatment**

These treatments involve processes where strictly physical phenomena are involved in the treatment of water. *Coagulation-flocculation* is a technique that uses the difference in density between the water and the suspended material for the chemical treatment of wastewater prior to sedimentation and filtration.[42] Coagulation typically involves the addition of certain substances which facilitate the removal of suspended particles. These can either be clay and silt, or organic material such as algae, bacteria, viruses, protozoa and natural organic matter. The generally used inorganic coagulants are aluminium sulfate (or alum) or iron(III) salts such as iron(III) chloride that are used to neutralize charges and form a gelatinous mass to trap particles. Subsequently, the mixture is gently stirred or agitated to generate larger particles or "flocs" so that they can be easily settled or filtered from the solution and this process is termed "flocculation." This step is then followed by sedimentation wherein waste water is allowed to stand for a short period of time in a tank under quiescent conditions due to which the heavier solids settle down under the force of gravity. Finally, the water is filtered using either rapid sand filters or membrane filters to remove remaining suspended particles or unsettled floc (Figure 2.7).

(3) **Aerobic/anaerobic biological treatment**

Biological treatment methods that have often been employed for dealing with pharmaceutical waste water traditionally use microorganisms, particularly bacteria, for the decomposition of contaminants present in the effluent. Both organic as well as inorganic solids can be removed by this process. On the basis of the availability of dissolved oxygen, this technique can be further divided into two

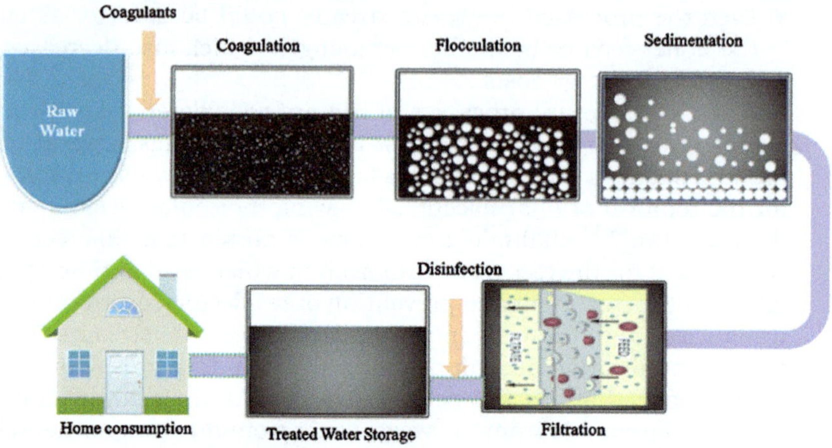

Figure 2.7 Physico-chemical treatment methodology for treatment of waste water.

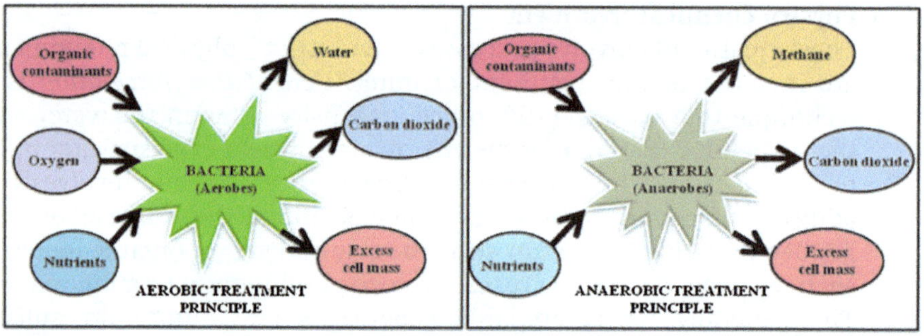

Figure 2.8 Aerobic and anaerobic treatment principles.

categories, namely: (i) aerobic treatment systems; and (ii) anaerobic treatment systems (Figure 2.8). The technology applied in the aerobic biological treatment includes the activated sludge process, AS with granular activated carbon, extended aeration activated sludge process, membrane bioreactors *etc*. As the name suggests, "aerobic treatment processes" take place in the presence of air wherein microorganisms also called "aerobes" utilize molecular or free oxygen for the decomposition of organic impurities and convert them in to carbon dioxide, water and biomass.[43–45] On the other hand, in anaerobic treatment processes, the assimilation of organic impurities occurs in the absence of air by the anaerobes to form methane, carbon dioxide and biomass.[46–48]

(4) **Advanced oxidation processes**
Several pharmaceuticals that have low biodegradability cannot be completely removed from the waste water by the commonly employed water purification methodologies and therefore, in such cases,

advanced oxidation processes (AOP) are used.[49] AOPs can be broadly defined as aqueous-phase oxidation methods that cause the destruction of harmful pollutants *via* oxidation through reactions with hydroxyl radicals.[50–55] Depending upon the nature of the pharmaceutical effluent, different types of AOPs may be employed:

- Heterogeneous and homogeneous photocatalysis
- Electro-oxidation
- Fenton and photo-Fenton process
- Wet air oxidation
- Ozone/H_2O_2 (peroxide) treatment
- Microwave and ultrasonic irradiation

The basic mechanism involved in AOPs can be summarized in Figure 2.9. The first step involves the formation of strong oxidants *i.e.* hydroxyl radicals. In the next step, oxidants react with the organic contaminants present in the water to generate biodegradable intermediates. Finally, the biodegradable intermediates react with oxidants in a process referred to as mineralisation (*i.e.* production of water, carbon dioxide and inorganic salts).

(5) **Hybrid technologies**
None of the single treatment technologies are efficient enough for the complete eradication of toxic pharmaceutical contaminants from various water sources. It is here where the crucial role of hybrid technologies – *"combinations of one or more conventional or advanced treatment technologies"* becomes clear. Hybrid technologies have been successfully utilized for the treatment of chemical synthesis process waste water containing copious concentrations of organic pollutants ranging from reagents to intermediates to the final products. For example, Chen *et al.* used a combination of a two-phase anaerobic digestion (TPAD) system and a membrane bioreactor (MBR) which showed efficacy in removing 99% chemical oxygen demand (COD) from chemical synthesis-based waste water while simultaneously reducing the pH to the neutral range.[56] Boroski *et al.* on the other hand

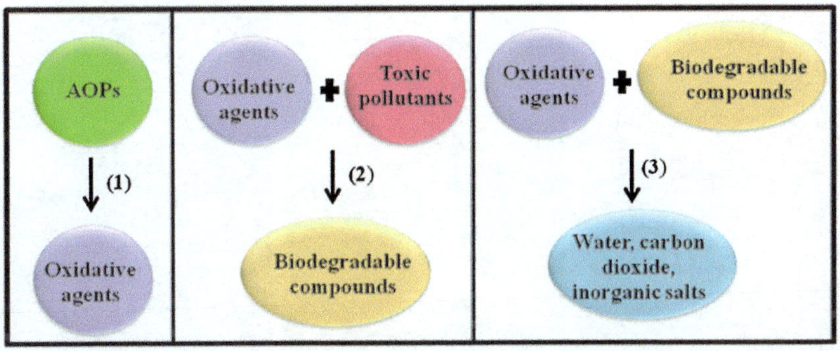

Figure 2.9 Steps involved in AOPs treatment of wastewater containing toxic organic compounds.

could achieve a removal efficiency of almost 86% COD and 90% turbidity from waste water containing large concentrations of non-biodegradable chemicals *via* the use of the electrocoagulation methodology and heterogeneous photocatalysis.[57] Apart from these, hybrid technologies involving the combination of pretreatment ozonation and MBR have also proven to be useful for the complete treatment of API formulation wastes comprising estrogens or penicillin formulation wastes containing harmful pollutants such as tylosin that have a refractory action on biological processes.

2.5.2 Case studies

2.5.2.1 AstraZeneca

Beginning its journey in the year 1953, AstraZeneca, an Anglo-Swedish multinational biopharmaceutical company has now achieved fame as a world class manufacturer of generic drugs for the treatment of several serious disorders *viz.* cancer, cardiovascular, gastrointestinal, infection, neuro, respiratory, and inflammation. To address today's pressing global challenges, the company has currently adopted the sustainability strategy "*Securing our Future*" deeper into their core business with a particular emphasis on the reduction of greenhouse gas emissions, waste water reuse and minimizing the environmental impact of our products. Within this framework, AstraZeneca has developed an effluent plant at Avlon Works, Avonmouth, United Kingdom to meet current and possible future environmental legislation, safeguard the company's assets and protect the coastal waters of the estuary.[58] The plant provides on-site treatment of aqueous effluent that was previously being sent out for treatment to the local coastal waters under consent from the Environment Agency. The plant produces four effluent streams as elaborated in Figure 2.10.

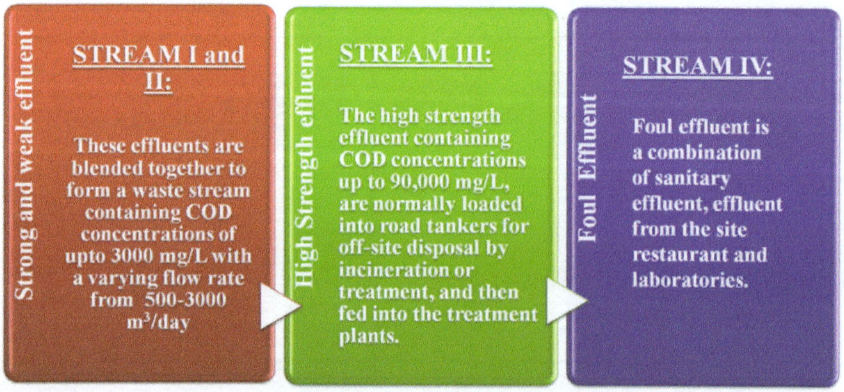

Figure 2.10 Four effluent streams produced by the plant.

The developed effluent treatment plant is designed to receive all four above mentioned streams with a total flow rate ranging from 500–2500 $m^3\,day^{-1}$ with a COD of approx. 3000 $mg\,L^{-1}$. The effluent treatment plant has a capacity of 4200 $Kg\,d^{-1}$ COD. The effluent treatment process is a Moving-Bed Bioreactor (MBBR) followed by chemical precipitation of orthophosphate, flocculation and dissolved air flotation (DAF).

2.5.2.1.1. Process Description. *Preliminary treatment consists of:*

- Coarse solids separation
- Blending and balancing
- Nutrient addition and pH balancing

Secondary treatment consists of:

- Biological treatment utilizing the Anox suspended carrier aeration process in two parallel reactor streams each comprising three reactor tanks
- Storage of final effluent during periods of low tide

Sludge treatment consists of:

- Treatment to remove and thicken sludge produced in the preliminary and secondary treatment processes

The complete schematic showing the AstraZeneca effluent plant at Avlon Works, Avonmouth, United Kingdom has been presented in Figure 2.11.

2.5.2.2 *Pfizer*

Established in 1849 by Charles Pfizer and his cousin Charles Erhart, Pfizer is one of the largest global pharmaceutical companies and is engaged in the discovery, development, manufacturing and marketing of prescription drugs and medicines. Within a very short time span, the annual worldwide sales of this company grew to $500 million. In the year 2003, Pfizer was included as a member of World Business Council for Sustainable Development (WBCSD) and business for social responsibility (BSR) organizations that provide resources to world leading corporations and concerned stakeholders for promoting sustainable business practices internationally. With the fast-growing popularity of sustainable business strategy, Pfizer has been looking for green innovative pathways for the formulation of APIs to minimize toxicological impact on environment. For example, Pfizer won Crystal Faraday Award for its contribution towards the minimization of environmental footprint of the synthetic process of the active ingredient of Viagra by reducing the amount of solvent waste generated from 1300 to 22 per kilogram of the product.[59]

The recovery and subsequent reuse of organic solvents is another ingenious approach that has conquered new horizons in the field of medicine and drug discovery. Studies reveal that reusing solvents instead of incinerating them is a much more sustainable and greener solution that is not only

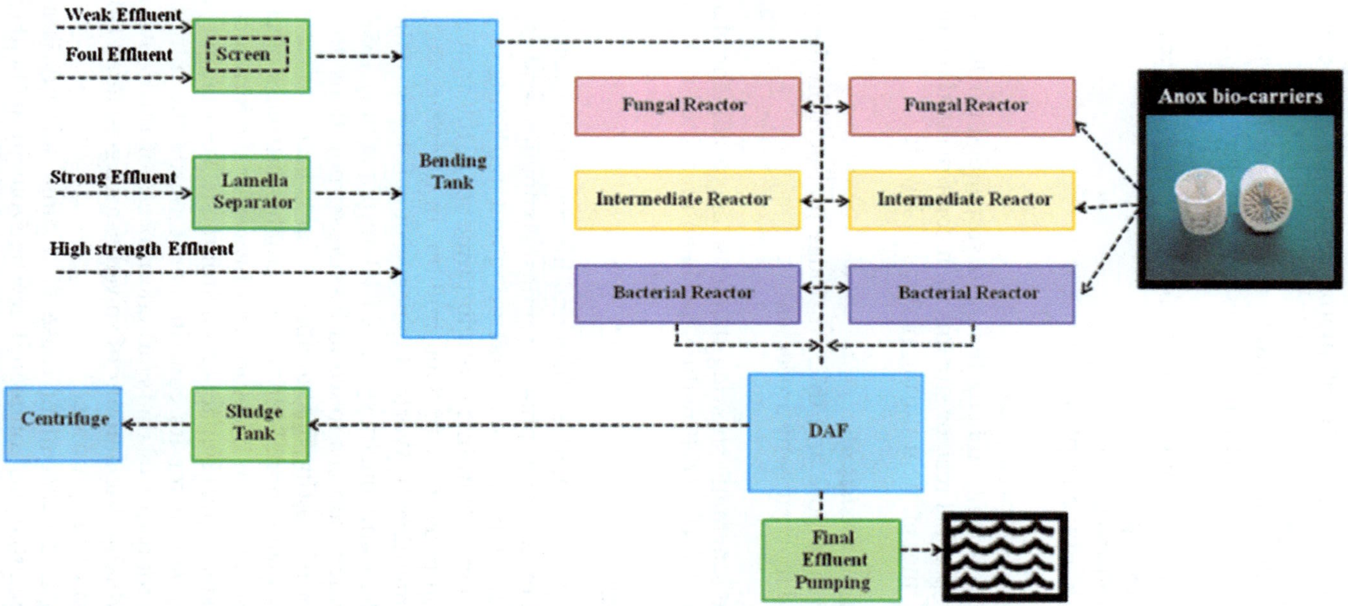

Figure 2.11 Treatment scheme for AstraZeneca effluent plant at Avlon Works, Avonmouth, United Kingdom.

important from an ecological viewpoint but also contains a real economic benefit for the industrial user. In order to meet the rising demands of energy and sustainability, Pfizer designed an alternative synthetic approach for the manufacture of sertraline hydrochloride – an active ingredient of Zoloft which is the most widely prescribed medication used to treat clinical depression.[60]

In the traditional manufacturing process, Stobbe's reaction was utilized to couple benzophenone with diethylsuccinate for the generation of a mono acid. Subsequently, the mono acid was hydrolysed and decarboxylated under strong acidic conditions to produce butenoic acid which upon hydrogenation with a Pd/C catalyst furnished 4,4-diarylbutanoic acid. In the next step, Friedel Crafts acylation and cyclization processes were applied to obtain the racemic tetralone intermediate which acted as the starting material for the target product. The tetralone intermediate thus obtained was reacted with methylamine to produce imine that was further hydrogenated to generate the desired sertraline hydrochloride. The reaction involved in the formation of imine is reversible and hence $TiCl_4$ was required to drive the equilibrium to the desired product by eliminating water.

However, the use of $TiCl_4$ led to the generation of several toxic by-products such as hydrochloric acid, titanium oxychloride, and titanium dioxide, the disposal of which required an additional process (energy), inputs (washes and neutralizers), and costs (waste disposal). The process efficiency team of Pfizer discovered that the formation of imine could be pushed to greater than 95% completion by substituting the solvent from THF to ethanol. As $TiCl_4$ was no longer needed for the imine formation, the new process alleviated environmental issues associated with transport, handling, and disposal of titanium wastes. The new process not only led to a significant reduction in the usage of one of the starting materials *i.e.* methyl methacrylate but also allowed it to be recycled back into the process with much more efficiency (Scheme 2.2). Another fruitful outcome of this modified approach was the implementation of a $Pd/CaCO_3$ catalyst that resulted in better region-selectivity than the existing catalytic system Pd/C. Besides, no unwanted products were generated using the more selective palladium catalyst which substantially reduced the need for huge amounts of solvent as well as energy. Most importantly, this process allowed the recovery and recycling of the catalyst that in addition to increasing the overall efficiency also addressed the economic and environmental concerns.

Another case study on using pervaporation (PV) for the recovery of isopropanol solvent in the manufacturing of **celecoxib**, the active ingredient in Celebrex® drug has been performed by Rowan University in collaboration with Pfizer.[61] The manufacturing operation at their plant in Barceloneta, Puerto Rico was assessed and it was found that during the final crystallization and production step, large quantities of waste were generated which included approximately equal amounts of isopropanol (IPA) and water, small amounts of methanol, ethanol and other dissolved solids. So, the prime goal was the purification and recovery of IPA from the waste streams.

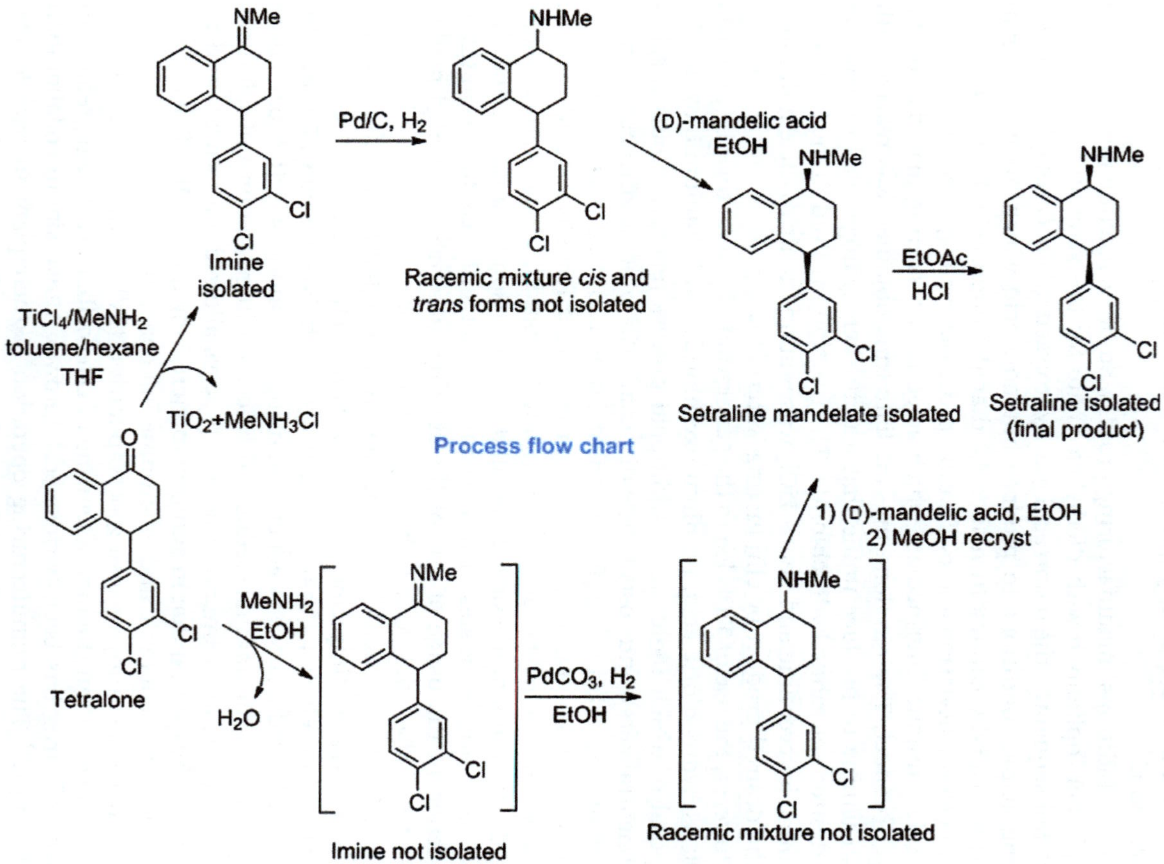

Scheme 2.2 Modified approach adopted by Pfizer for the manufacture of sertraline hydrochloride – an active ingredient of Zoloft.

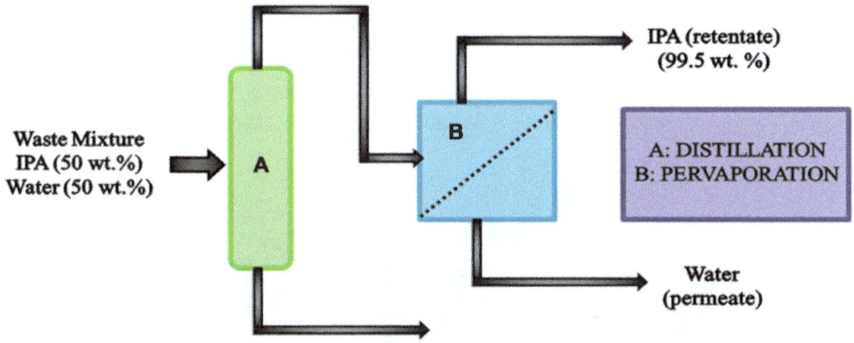

Figure 2.12 IPA recovery scheme in the celecoxib case.

A separation scheme based on distillation-PV-distillation system was therefore proposed to produce IPA at high enough purity to be recycled. This process led to a large-scale reduction in overall operating cost for the plant (almost 72%) when IPA purchase costs and waste disposal costs were analysed (Figure 2.12).

To enhance the economic feasibility of potential capital investment and improve the environmental footprint further, the Rowan team also evaluated the synthetic procedure for selamectin (an active ingredient in the drug Revolution®) which is being generated at the Pfizer Kalamazoo plant.[62] Considering the relatively high cost of the solvent acetonitrile used during selamectin synthesis, the acetone–acetonitrile waste stream was selected for this conceptual study. For the recovery of acetonitrile, a small-scale distillation solvent recovery operation unit was designed which proved to be a better alternative in comparison to the current waste disposal practice (Figure 2.13).

Besides, they analysed other binary waste streams (THF-Water and toluene-acetone) from the same plant and designed a simple solvent distillation unit to recover IPA used during the manufacture of nelfinavir, the active ingredient in the antiretroviral drug **"Viracept"** and toluene used during hydrocortisone manufacture (used in several drug products for relief of inflammation) (Figure 2.14).

In addition, they analysed other binary waste streams such as IPA-THF used during manufacture of nelfinavir, and toluene-acetone used during hydrocortisone manufacture in the same plant and designed simple solvent distillation units to separate and recover solvents (Figure 2.15).

2.5.2.3 Bristol-Myers Squibb

Bristol-Myers Squibb (BMS), an American biopharmaceutical company headquartered in New York city, has focused on the development of therapeutic drugs that help millions of people fight against deadly diseases such as cancer, HIV/AIDS, cardiovascular disease, diabetes, rheumatoid

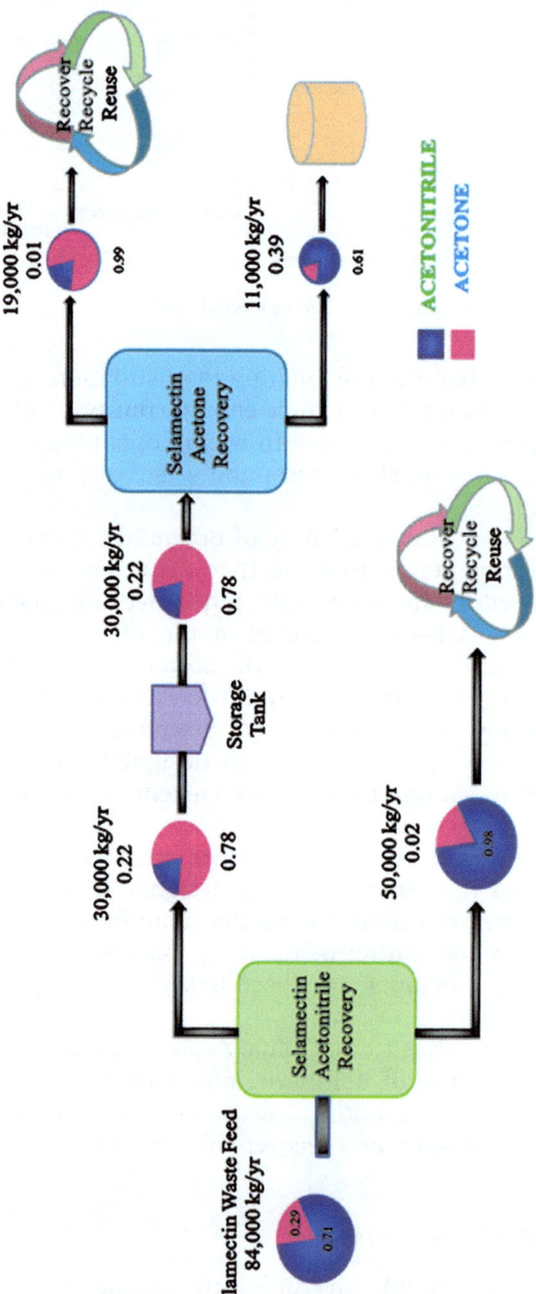

Figure 2.13 Process scheme for solvent recovery.

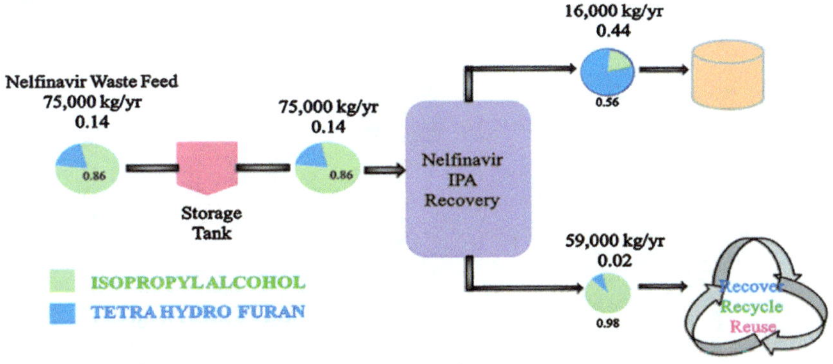

Figure 2.14 IPA recovery scheme in the nelfinavir case.

arthritis and many more.[63] In order to expand impact and profit, BMS also integrated some of the important green chemistry principles like waste minimization, substituting solvents, applying enzymes instead of chemicals and recycling of reagents into the company's operations. For instance, the BMS scientists and engineers developed a PV membrane technology for recovering solvents within a pilot-scale facility during the production of a new oncology drug which was a dramatic breakthrough in the pharmaceutical sector.[64] During one step, a mixture of organic solvent (THF), water and a pharmaceutical intermediate in a batch reactor must be dehydrated so that the water content gets reduced to 0.5% before further processing. For carrying out the dehydration step, a batch constant volume distillation (CVD) process, often termed as a "Dean Stark process," is employed which requires significant amounts of entrainer *i.e.* pure THF to break the azeotrope (THF-water mixture has an azeotrope at 95.7% THF at STP) and allow the batch water content to be reduced to the desired concentration. This process is not green as it leads to generation of large quantities of waste due to the usage of additional entrainers and other chemicals. Therefore, a proposal was made in which the PV technology was effectively combined with the batch CVD process to design CVD-PV green drying hybrid system (Figure 2.16).

In this technique, the role of the PV membrane is to dehydrate THF to the desired purity and recycle it back to the CVD vessel. This membrane process not only reduces the waste disposal cost by 93% but also cuts down the cost of purchase of THF by 56%. Besides, it also avoids the greenhouse gas emissions through solvent purification and recycling that would address the growing threat of climate change.

2.5.2.4 ChemGenes Corporation

ChemGenes Corporation is a biotechnology company located in Wilmington, M.A which deals with raw materials or building blocks for RNA/DNA

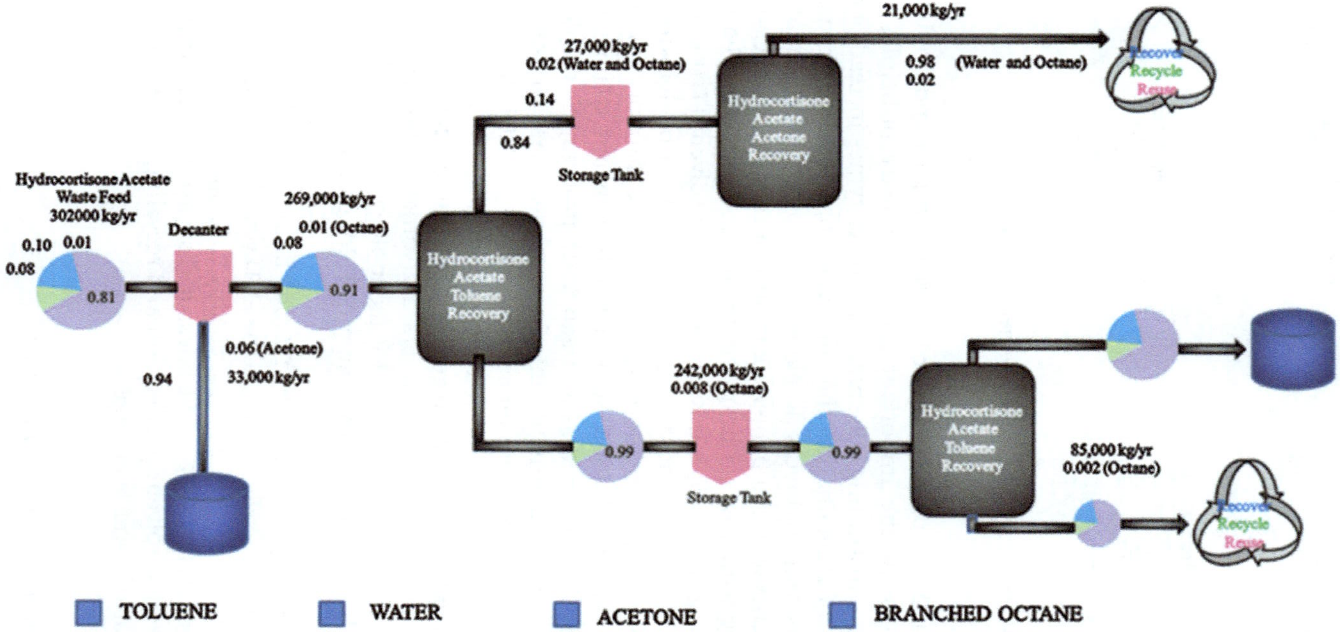

Figure 2.15 Simple solvent distillation units to separate and recover solvents used during the manufacture of nelfinavir.

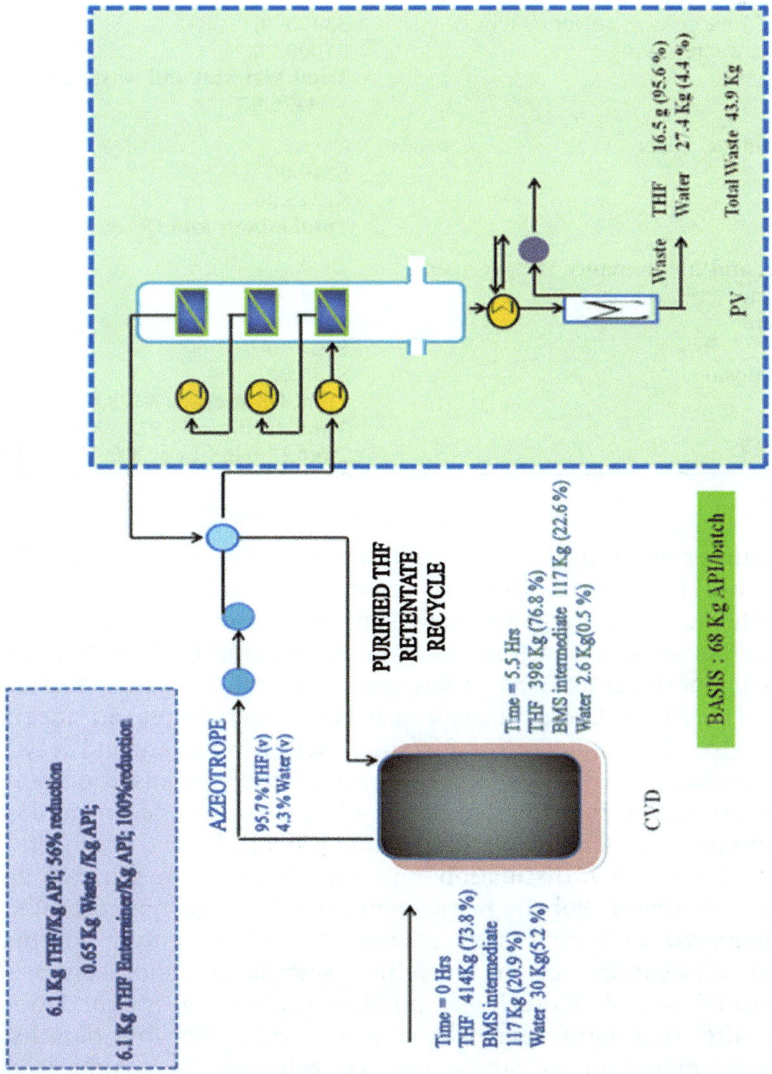

Figure 2.16 Designed CVD-PV green drying hybrid system.

Table 2.4 Results obtained with the TechnoClean F-2500 within four months.

Usage and costs for four months	Quantity/costs
Input solvent	1339 L
Output solvent	1308 L
% Recovery	97.70%
Material and waste savings	
Within 97.70% return, solvent savings	$3475.76
Hazardous waste saving	$1000.90
	Total Material and waste savings: $4476.65
Labour and QC costs	
Labour	$810.00
QC	$1215.00
	Total labour and QC costs: $2025.00
Operation and maintenance (O&M) costs	
Maintenance	$214.65
Tank liners	$21.60
Electricity	$107.00
Waste disposal	$135.00
	Total O&M costs: $478.45
	Total savings: $1,973.22
	Over 27 batches or $73.27 per batch

manufacturing often used in academic research. Since 2005, ChemGenes has been working with the Office of Technical Assistance and Technology (OTA) with the aim of reducing toxic solvent consumption and managed to reduce the use of chloroform by 55% and hexane by 35% that results in net saving of dollar $215 000. A few years later, the Toxics Use Reduction Institute (TURI) at UMass Lowell provided ChemGenes an incentive grant to help offset capital costs for a new solvent recovery and recycling system. Encouraged by the success of various projects based on solvent reduction strategies, they focused on recycling blends of hexane and ethyl acetate produced during the manufacturing process. With this objective in mind, at first a 100 L distillation unit was designed to separate spent solvent for reusing it. But the instrument operation was not cost effective when compared to buying fresh solvent. Therefore, a fully automated fractional distillation solvent recycling system, TechnoClean F-2500 manufactured by CBG Biotech was purchased under recommendation of the OTA. After thoroughly analysing this new process for four months, an overall cost reduction by 50% could be achieved as summarized in Table 2.4.

2.5.2.5 Novartis

Novartis is a world leader pharmaceutical corporation in the research and development of therapeutic drugs that was created in the year 1996 through

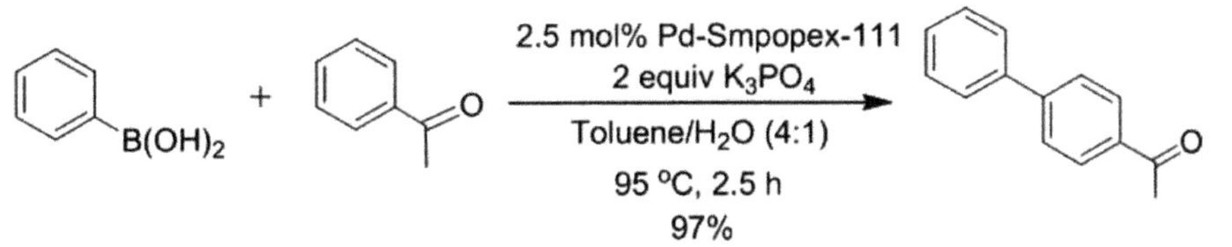

Scheme 2.3 Suzuki cross-coupling of phenyl boronic acid and bromide in the presence of the palladium Smopex catalyst.

the merger of two companies having a rich and diverse corporate history, namely Ciba–Geigy and Sandoz. It has been recognized worldwide for its admirable reputation as it offers high quality prescription drugs that address a wide range of ailments which include cardiovascular diseases, central nervous system disorders, cancers, organ transplant complications *etc.* One of the key objectives of the **Novartis Pharmaceutical Corporation** is to discover new strategies for improving the current synthetic protocols for waste minimization. To address the separation and contamination issues associated with homogeneous catalysis, Novartis scientists designed a new heterogeneous Pd catalytic system through immobilization of palladium onto a Smopex-111 support matrix (a-fibre based benzenethiol) for Suzuki Coupling reaction.[65] This Smopex-based catalyst not only exhibited excellent activity in the concerned coupling reaction but also allowed effective recovery of precious palladium metal from the process stream. As evident from the scheme, a nearly quantitative yield of the desired coupling product has been obtained in Suzuki cross-coupling of phenyl boronic acid and bromide in presence of the developed palladium Smopex catalyst. The most striking feature of this novel route was that the catalyst could be recycled up to four cycles without any discernible loss in its activity (Scheme 2.3).

2.5.2.6 GlaxoSmithKline

GlaxoSmithKline (GSK), a science-led global healthcare company with a market capitalisation value of £81, the fourth largest on London Stock Exchange works with the mission to develop a broad range of innovative products in the area of pharmaceuticals. Amongst a wide variety of consumer goods produced by GSK, cost effective and innovative respiratory inhalers are the top selling products of this company which are generally used to treat asthma and chronic lung diseases. However, when they began to look into the carbon footprint in the year 2010, it became quite clear that the utilization of these respiratory inhalers adversely affected the environment due to production of 5.2 million tonnes of CO_2 emissions. An in-depth analysis of the cause behind the greenhouse gas emissions revealed that 95% of the footprint generated by metered dose inhalers (MDIs) was a result of the propellant used for their fabrication while 65% of the footprint of the Diskus dry-powder inhalers (which do not need propellants) came from the plastics and the energy used in its manufacture. In order to combat this problem, they have launched the first recycling and recovery scheme – "Complete the cycle" for breathing new life into all respiratory inhalers. Under this scheme, following steps have been taken to reduce environmental impact of landfill disposal of respiratory inhalers:

1. Firstly, all the inhalers were collected as a part of routine deliveries and sent to a waste management company (specialized recyclers) to be re-sorted for recovery.
2. The use of plastic has been replaced by aluminium in the inhaler devices as aluminium can be recycled easily for further use.
3. By switching over from CFC to HFA propellant gases, GSK has been able to reduce the emissions from its respiratory inhalers by 80% per pack.

2.5.2.7 Merck

Ranking number one on a recent list of the top 1000 sustainable companies is Merck, a recipient of the United States EPA Energy Star Sustained Excellence Award, has been actively involved in the design and development of recyclable materials with reduced life cycle impacts. As part of their commitment towards reducing the environmental impacts of manufactured products, Merck, in collaboration with five customers, jointly launched a pilot recycling program in United States in the year 2012. Under this scheme, over 189 metric tonnes of single-use plastic products were diverted from landfill and collected, out of which they were able to recycle 91 metric tonnes of plastic. An end-of-life cycle assessment revealed that this pilot project could reduce greenhouse gas emissions by up to 14% compared to traditional disposal methods.

llinois EPA Recommended Handling/Disposal Options for Households:[66]

1. **Reduce** pharmaceutical waste whenever possible:
   - Use all antibiotics as prescribed by your doctor. Not using all doses of an antibiotic could lead to the development of antibiotic resistant strains of bacteria.
   - Buy only as much as can reasonably be used before the expiration date. For example, do not buy 500 aspirin just because it is cheaper, unless you will use them all.
   - When your doctor prescribes a new medication, ask the doctor to prescribe only enough to see if the medication will work for you and in the lowest dose advisable. That way, if the medication does not suit you, there is less to waste. Do the same for your pet's medications.
2. **Dispose** of the remainder properly:
   - Whenever possible, take your unused pharmaceuticals to a pharmaceutical collection program or household hazardous waste collection event. If a collection program does not exist in your area,

encourage your health care provider, local governments and local law enforcement to develop one. **Note**: If you choose to store your waste while you wait for a pharmaceutical collection event to occur, please minimize the risk of accidental poisoning, overdose or diversion (illegal use by someone other than the intended person) by storing medications out of reach of children or in a locked cabinet.

- Dispose of your unused pharmaceuticals in the trash. Especially when there is a risk of accidental poisoning, overdose or diversion, it is better to dispose of household pharmaceuticals than to hang onto them. When placing unused pharmaceuticals in the trash, be sure to do the following:
  - remove or mark over all labels that identify the materials as pharmaceuticals or that could provide personal information about you, including prescription information that someone could try to refill;
  - render them unattractive to children and thieves by dissolving them in a small amount of water/alcohol or by grinding them up and mixing them with coffee grounds or kitty litter; and
  - put them in a second container or small plastic bag and hide them in your trash.
- Never burn pharmaceuticals or personal care products in a burn barrel. Uncontrolled burning can create dioxins and other air pollutants.

2.6 Conclusion

Although the pharmaceutical sector has played a significant role in improving the quality of our lives, the disposal of drugs into water supplies, rivers, lakes and other waterways have put pharmaceutical chain supplies under spotlight. As a consequence, a great array of endeavours of industrial researchers and engineers are now being directed towards the redesigning of existing synthetic protocols with the aid of green chemistry principles and practices such as the elimination of hazardous substances, reuse and recycling of reagents, use of biosynthetic routes and solvent substitution practice. In this context, it is worth mentioning that substantial achievements have already been made by several renowned companies like Pfizer, GSK, Novartis *etc.* as they have come up with some brilliant recovery and recycling strategies, yet collective efforts of environmentalists, regulators, manufacturers and the public are still required to find solutions to the global environmental problem of hazardous waste management.

References

1. T. Ishibashi, M. Kusama, Y. Sugiyama and S. Ono, *J. Clin. Pharm. Ther.*, 2012, **37**, 657–663.
2. A. Bauer and M. Bronstrup, *Nat. Prod. Rep.*, 2014, **31**, 35–60.

3. K. G. Kohlstaedt, *Clin. Pharmacol Ther.*, 1960, **1**, 192–201.
4. S. R. Hughes, P. Kay and L. E. Brown, *Environ. Sci. Technol.*, 2013, **47**, 661–677.
5. M. Agerstrand, C. Berg, B. Björlenius, M. Breitholtz, B. Brunström, J. Fick, L. Gunnarsson, D. G. J. Larsson, J. P. Sumpter, M. Tysklind and C. Rudén, *Environ. Sci. Technol.*, 2015, **49**(9), 5336–5345.
6. K. Sreekanth, N. V. Gupta, H. V. Raghunandan and U. N. Kashyap, *Int. J. PharmTech. Res.*, 2014, **6**(3), 899–907.
7. B. Kasprzyk-Hordern, *Chem. Soc. Rev.*, 2010, **39**, 4466–4503.
8. K. A. Heys, R. F. Shore, M. G. Pereira, K. C. Jones and F. L. Martin, *RSC Adv.*, 2016, **6**, 47844–47857.
9. D. G. J. Larsson, *Philos. Trans. R. Soc., B*, 2014, **369**, 1656–1664.
10. B. W. Cue and J. Zhang, *Green Chem. Lett. Rev.*, 2009, **2**(4), 193–211.
11. R. Bandichhor, A. Bhattacharya, A. Cosbie, L. Diorazio, P. Dunn, K. Fraunhoffer, F. Gallou, J. Hayler, M. Hickey, B. Hinkley, L. Humphreys, B. Kaptein, L. Oh, P. Richardson, S. Roberts, T. White, S. Wuyts and J. Yin, *Org. Process Res. Dev.*, 2015, **19**(12), 1924–1935.
12. C. Gadipelly, A. Pérez-González, G. D. Yadav, I. Ortiz, R. Ibáñez, V. K. Rathod and K. V. Marathe, *Ind. Eng. Chem. Res.*, 2014, **53**(29), 11571–11592.
13. S. K. Khetan and T. J. Collins, *Chem. Rev.*, 2007, **107**(6), 2319–2364.
14. K. Sanderson, *Nature*, 2011, **469**, 18–20.
15. P. Gupta and A. Mahajan, *RSC Adv.*, 2015, **5**, 26686–26705.
16. C. R. McElroy, A. Constantinou, L. C. Jones, L. Summerton and J. H. Clark, *Green Chem.*, 2015, **17**, 3111–3121.
17. C. Teixeira, N. Vale, B. Pérez, A. Gomes, J. R. B. Gomes and P. Gomes, *Chem. Rev.*, 2014, **114**, 11164–11220.
18. M. Koel, *Green Chem.*, 2016, **18**, 923–931.
19. R. A. Sheldon, *Green Chem.*, 2016, **18**, 3180–3183.
20. S. C. DeVito, *Green Chem.*, 2016, **18**, 4332–4347.
21. C. G. Daughton, *Sci Total Environ.*, 2016, **562**, 391–426.
22. D. Kapoor, *J. Crit. Rev.*, 2015, **2**, 25–30.
23. K. Grodowska and A. Parczewski, *Acta Pol. Pharm.*, 2010, **67**, 3–12.
24. D. J. C. Constable, C. Jimenez-Gonzalez and R. K. Henderson, *Org. Process Res. Dev.*, 2007, **11**, 133–137.
25. H. Federsel, *Green Chem.*, 2013, **15**, 3105–3115.
26. A. D. Curzons, D. C. Constable and V. L. Cunningham, *Clean Prod. Process*, 1999, **1**, 82–90.
27. C. A. Busacca, D. R. Fandrick, J. J. Song and C. H. Senanayake, *Adv. Synth. Catal.*, 2011, **353**, 1825–1864.
28. H. Doucet and J. C. Hierso, *Curr. Opin. Drug Discovery Dev.*, 2007, **10**, 672–690.
29. I. V. Gürsel, T. Noël, Q. Wang and V. Hessel, *Green Chem.*, 2015, **17**, 2012–2026.
30. P. Marchetti, M. F. J. Solomon, G. Szekely and A. G. Livingston, *Chem. Rev.*, 2014, **114**, 10735–10806.
31. A. Dastan, A. Kulkarni and B. Torok, *Green Chem.*, 2012, **14**, 17–37.

32. E. Rafiee and S. Eavani, *RSC Adv.*, 2016, **6**, 46433–46466.
33. D. J. Constable, A. D. Curzons and V. L. Cunningham, *Green Chem.*, 2002, **4**, 521–527.
34. C. Gadipelly, A. Pérez-González, G. D. Yadav, I. Ortiz, R. Ibáñez, V. K. Rathod and K. V. Marathe, *Ind. Eng. Chem. Res.*, 2014, **53**, 11571–11592.
35. K. A. Kidd, P. J. Blanchfield, K. H. Mills, V. P. Palace, R. E. Evans, J. M. Lazorchak and R. W. Flick, *Proc. Natl. Acad. Sci. U. S. A.*, 2007, **104**, 8897–8901.
36. J. Weinberger and R. Klaper, *Aquat. Toxicol.*, 2014, **151**, 77–83.
37. M. Hampel, J. E. Bron, J. B. Taggart and M. J. Leaver, *Aquat. Toxicol.*, 2014, **151**, 114–123.
38. M. Sun, S. X. Gan, D. F. Yin, H. Y. Liu and W. D. Yang, *Chin. J. Antibiot.*, 2000, **25**, 172–174.
39. W. Zhang, G. He, P. Gao and G. Chen, *Sep. Purif. Technol.*, 2003, **30**, 27–35.
40. A. Shahtalebi, M. H. Sarrafzadeh and M. M. Rahmati, *Iran. J. Environ. Health Sci. Eng.*, 2011, **8**, 109.
41. J. Bezawada, S. Yan, R. P. John, R. D. Tyagi and R. Y. Surampalli, *Biotechnol. Res. Int.*, 2011, 1–11.
42. B. V. Reddy, P. Sandeep, P. Ujwala, K. Navaneetha and K. V. R. Reddy, *Water Treatment Process in Pharma Industry-A Review*, 2014, vol. 4, pp. 7–18.
43. D. S. S. Raj and Y. Anjaneyulu, *Process Biochem.*, 2005, **40**, 165–175.
44. T. M. LaPara, C. H. Nakatsu, L. M. Pantea and J. E. Alleman, *Water Res.*, 2001, **35**, 4417–4425.
45. M. Z. H. Khan and M. G. Mostafa, *Int. J. Environ. Sci.*, 2011, **1**, 1797.
46. S. Yuzir, M. M. Din and P. J. Sallis, *Int. J. Chem. Eng. Appl.*, 2011, **2**, 32.
47. Y. A. Oktem, O. Ince, P. Sallis, T. Donnelly and B. K. Ince, *Bioresour. Technol.*, 2008, **99**, 1089–1096.
48. A. G. Rao, G. V. Naidu, K. K. Prasad, N. C. Rao, S. V. Mohan, A. Jetty and P. N. Sarma, *Bioresour. Technol.*, 2005, **96**, 87–93.
49. K. Kümmerer, *Pharmaceuticals in the Environment: Sources, Fate, Effects and Risks*, Springer Science & Business Media, 2008.
50. A. Vogelpohl and S. M. Kim, *Ind. Eng. Chem. Res.*, 2004, **10**, 33–40.
51. M. Klavarioti, D. Mantzavinos and D. Kassinos, *Environ. Int.*, 2009, **35**, 402–417.
52. C. Comninellis, A. Kapalka, S. Malato, S. A. Parsons, I. Poulios and D. Mantzavinos, *J. Chem. Tech. Biotechnol.*, 2008, **83**, 769–776.
53. I. A. Balcıoğlu and M. Ötker, *Chemosphere*, 2003, **50**, 85–95.
54. J. R. Domínguez, T. González, P. Palo, J. Sánchez-Martín, M. A. Rodrigo and C. Sáez, *Water Air Soil Pollut.*, 2012, **223**, 2685–2694.
55. I. A. Alaton, S. Dogruel, E. Baykal and G. Gerone, *J. Environ. Manage.*, 2004, **73**, 155–163.
56. Z. Chen, N. Ren, A. Wang, Z. P. Zhang and Y. Shi, *Water Res.*, 2008, **42**, 3385–3392.

57. M. Boroski, A. C. Rodrigues, J. C. Garcia, L. C. Sampaio, J. Nozaki and N. Hioka, *J. Hazard. Mater.*, 2009, **162**, 448–454.
58. P. B. Joshi, *J. Chem. Chem. Sci.*, 2016, **6**, 989–996.
59. P. J. Dunn, A. Wells and M. T. Williams, *Green Chemistry in the Pharmaceutical Industry*, John Wiley & Sons, 2010.
60. R. M. Pinto and S. M. Silvestre, *J. Chem. Eng. Chem. Res*, 2014, **1**, 290–301.
61. C. S. Slater, M. Savelski, G. Hounsell, D. Pilipauskas and F. Urbanski, *Proceedings of American institute of Chemical Engineers Annual Meeting*, Philadelphia, PA, 2008.
62. M. J. Raymond, C. S. Slater and M. J. Savelski, *Green Chem.*, 2010, **12**, 1826–1834.
63. K. C. Nicolaou, R. K. Guy and P. Potier, *Sci. Am.*, 1996, **274**, 84–88.
64. C. S. Slater, M. J. Savelski, T. M. Moroz and M. J. Raymond, *Green Chem. Lett. Rev.*, 2012, **5**, 55–64.
65. T. Laird, *Org. Process Res. Dev.*, 2012, **16**, 1874.
66. C. Brown, M. Milke and E. Seville, *Waste Manage.*, 2012, **31**, 1085–1098.

CHAPTER 3

Recoverable Polymer-supported DMAP Derivatives

CHRISTOPHER E. HOBBS

Department of Chemistry, Texas A&M University-Kingsville,
700 University Blvd., Kingsville, TX 78363, USA
Email: christopher.hobbs@tamuk.edu

3.1 Introduction

The development of methods aimed at the replacement of potentially dangerous and toxic chemical reagents is a very important endeavor in modern organic chemistry. The inspiration of this is rooted in the emergence of "Green Chemistry".[1] Anastas described this philosophy as 12 operating principles that describe the importance of waste generation, using non-toxic and environmentally benign reagents, atom economy, and catalysis. These principles have been implemented into every branch of chemical sciences and have led to incredible advances in the way that chemists (especially synthetic chemists) create new compounds and materials.

Often, synthetic chemists are interested in utilizing the principles of Green Chemistry in ways that allow for the exclusion of toxic species and petroleum-derived organic solvents by developing transition-metal-free reactions or alternative reaction media, respectively.[2–8] However, in many cases potentially toxic substances can be used only if they are successfully sequestered away from reaction products.[9] As a consequence of this sequestration, these substances, in some cases, can be recovered and even recycled.[10] This is an especially important aspect with respect to catalysis

Green Chemistry Series No. 52
Hazardous Reagent Substitution: A Pharmaceutical Perspective
Edited by Rakesh Kumar Sharma and Rakeshwar Bandichhor

Published by the Royal Society of Chemistry, www.rsc.org

since many modern molecular catalysts are based on environmentally unfriendly transition metals.

This goal is often achieved by anchoring reagents, catalysts, or ligands to polymer supports.[11–13] This has allowed chemists the ability to exploit a certain polymer's unique solubility properties in order to successfully separate these hazardous species away from important products. This method also has the consequence of rendering processes more sustainable as it can dramatically reduce the amount of toxic solvents required for typical synthetic procedures, resulting in much lower levels of waste. This is an important issue for academics, but an especially important issue for industry considering the majority of waste generated is from solvent used in reactions carried out in process scales.[14]

3.2 History

Using polymers to recover, sequester, and reuse reagents and catalysts is rooted in the development of solid-phase synthesis which relied on cross-linked polystyrene (so-called Merrifield resins) for the preparation of peptides and oligonucleotides.[15] This work revolutionized the synthesis of biomolecules, which, up to this point, could be a daunting task. Because of this discovery by Merrifield and Letsinger,[16] biochemists and molecular biologists (who may have almost no synthetic experience) can successfully prepare complex organic compounds. This revolutionary idea led to the Nobel Prize being awarded to Merrifield in 1984.

This idea behind solid-phase synthesis takes advantage of the insolubility of the resins which can be used to pull growing peptide (or nucleotide) chains out of solution, affecting their separation from byproducts, unreacted starting material, or other impurities. From this came the idea of "heterogenizing" homogeneous catalysts by anchoring them to the same type of insoluble polymers.[17] This effectively allowed chemists the opportunity to rely on the same principle of using the resin's insolubility to separate a catalyst or reagent away from important products. This had the advantage of sequestration of potentially toxic species as well as decreasing the amount of solvent waste generated since wasteful work-ups and separations could be avoided. However, since the catalysts were rendered heterogeneous, they were insoluble before, during, and after the reaction. A common belief among chemists is that insoluble species can exhibit lower reactivity and selectivity, in comparison to their homogeneous counterparts. To circumvent this issue, the research group of Ernst Bayer in Germany was one of the earliest to utilize soluble polymers as supports for peptide synthesis as well as catalysis.[18,19]

3.3 Recovery Strategies of Soluble Polymer-supported Species

Recovery of soluble polymer-supported species is not as straightforward as it is for insoluble species (which requires little more than simple gravity filtration). Several methods have been developed to affect separation, sequestration and recycling. Many of these systems rely on exploiting of a specific polymer's solubility in different solvents. One of the methods Bayer utilized was a solvent precipitation in which linear polystyrene (PS)-supported homogeneous catalysts could be isolated upon the addition of a "poor" solvent, that is, one in which the polymer is insoluble.[18,19] Recovery of the supported catalyst relied on simple filtration. This principle has been applied immensely to poly(ethylene glycol) (PEG)-supported catalysts as well.[11] However, this system relies on the use of excess amounts of solvent that results in the generation of, in some cases, large amounts of waste. Although there are shortcomings, this precipitation method is still used today. Methods aimed at greener recoveries and isolations have been developed.

These methods can broadly be described as either solid–liquid or liquid–liquid separations (Figure 3.1). Bergbreiter's laboratory at Texas A&M University pioneered the use of polyethylene oligomers (PE_{olig})[11,12,20] as supports for catalysts/reagents under thermomorphic recovery systems (and this chemistry has recently reemerged[21–24]). Such systems are interesting as PE_{olig} have no solubility at room temperature in any organic solvent but, upon heating, completely dissolve in organic solvents such as toluene and THF (and various mixtures of solvents). So, PE_{olig}-supported catalysts or reagents can be used at high temperatures (>65 °C) and recovered as solids upon cooling back to room temperature (Figure 3.2). Although this idea has been applied in industry (du Pont[25] developed PE_{olig}-supported polymerization catalysts) the elevated temperatures may not be desired as they require excess energy input and can lead to catalyst/product decomposition or detrimentally affect selectivity.

This requirement for high temperatures and limited solubility can serve as an impediment for PE_{olig}-supported species. So many other supports and recovery strategies have been invented. Liquid–liquid[11–13] separations can be used with polymers that have high levels of phase-selective solubility for one solvent over the other. For example, nonpolar oligomers such as poly(4-dodecylstyrene),[26] poly(*N*-octadecylacrylamide),[27] polyisobutylene,[28] and

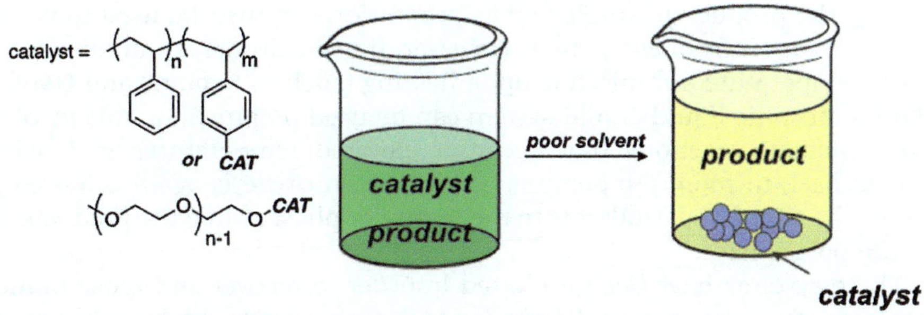

Figure 3.1 Precipitation recovery of PS or PEG-supported catalyst after a reaction.

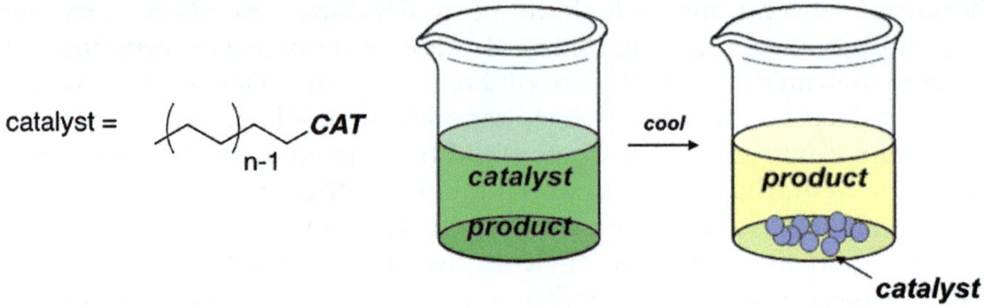

Figure 3.2 Thermomorphic recovery of PE_{olig}-supported catalyst after a reaction.

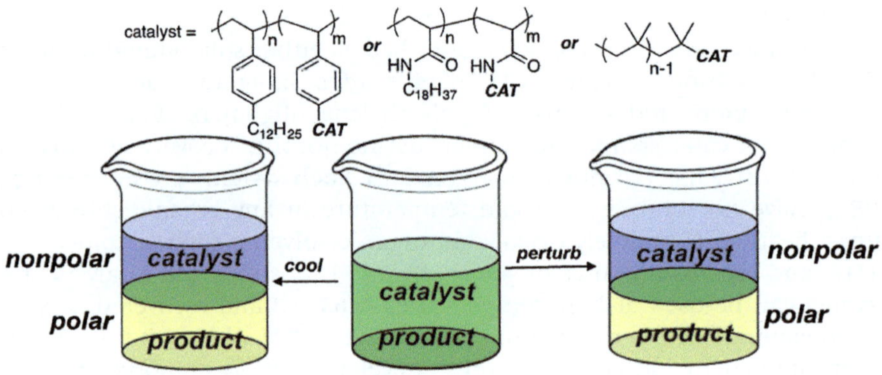

Figure 3.3 Liquid–liquid recoveries of polymer-supported catalyst after a reaction and some examples of polymers often used.

atactic-polypropylene[29] have been used as phase selective supports. Such materials can be used under both latent-biphasic and thermomorphic liquid–liquid systems. Latent-biphasic systems require reaction media to be composed of two (or more) solvents of opposing polarity that are on the cusp of miscibility at room temperature (such as heptane and ethanol). This insures that a supported catalyst can be used under homogenous reaction conditions. Upon reaction completion, phase separation is induced which results in a nonpolar phase (containing the catalyst) and polar phase (containing the products). Simple gravity separation can then be used to isolate the layers apart from each other. However, if two solvents are immiscible at room temperature but miscible upon heating (such as heptane and DMF), a thermomorphic liquid–liquid system can be used (Figure 3.3). This involves homogeneous reactions that occur at elevated temperatures and, when cooled back to room temperature, a biphasic mixture is again achieved in which the supported catalyst is in the nonpolar phase, while the products are in the polar phase.

These systems have been exploited in order to recover and reuse numerous catalysts, reagents, and ligands. This turns out to be a multi-faceted approach towards Green Chemistry as it allows for: (i) sequestration of

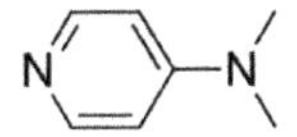

Figure 3.4 Structure of DMAP.

potentially toxic substances out of products; (ii) the recovery and reuse of catalysts/reagents: and (iii) the reduction of the carbon footprint that is commonly associated with chemical processes by reducing the amounts of solvent waste generated. The remainder of this chapter will focus on strategies that have been employed for the sequestration, recovery and reuse of *N,N*-dimethylaminopyridine (DMAP) (Figure 3.4) which is an organic base and organocatalyst that is ubiquitous in industry and academia.

3.4 DMAP

The growing impetus on green and sustainable chemistry has prompted the synthetic community to heavily explore methods and procedures that reduce and eliminate required, or generated, hazardous materials. Towards this goal, the field of transition metal-free catalysis (*organocatalysis*) has grown tremendously.[30] In the past couple of decades many metal-free catalysts have been developed and exploited for the construction of important compounds. Among these, nucleophilic amines are some of the oldest examples.

DMAP's prowess as a nucleophilic base and organocatalyst has long been known. DMAP has traditionally been used as a "super acylation"[31,32] catalyst for the esterification of alcohols and, more recently, the preparation of biologically relevant polyesters (Figure 3.5).[33] The use of DMAP (and its derivatives) as an alternative to traditional transition metal catalysis is spurred by the desire to rid processes of toxic and environmentally detrimental metal salts that can be inadvertently released into the ecosystem. However, ridding processes of transition metal salts is often not enough to curb the effect that chemical procedures can have on the environment.[9] This is because chemical processes rely heavily on organic solvents that must be carefully disposed of. In fact, the majority of industrial chemical waste is composed of solvent.[14] Furthermore, organocatalysts (like DMAP) can be highly toxic themselves (DMAP's LD_{50} can be as high as 450 mg kg^{-1}).[9] So, caution must be taken for their use and removal. For these reasons, the preparation of methods to recover and sequester toxic species is of the utmost importance. As stated in the earlier sections, this often achieved by using a polymer support.

3.4.1 DMAP Preparation

The synthesis of DMAP was first described approximately 90 years ago.[34] This method relied on reaction between 4-chloropicolinic acid and dimethylamine. Later, other methods (that allowed for its large-scale production) were described. DMAP could be obtained through reaction of

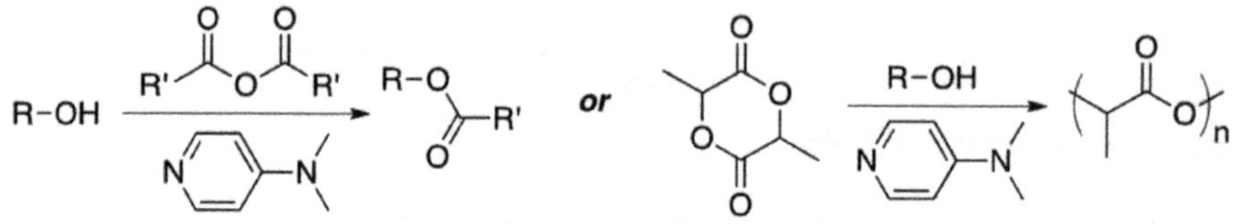

Figure 3.5 The use of DMAP as an acylation or polymerization catalyst.

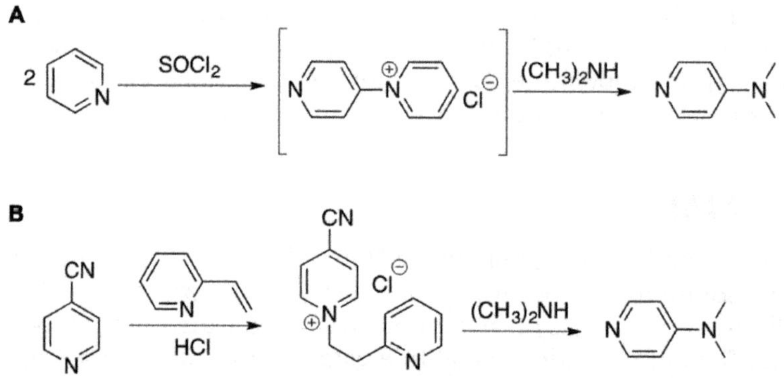

Figure 3.6 Syntheses of DMAP.

thionylchloride with two equivalents of pyridine, followed by reaction with dimethylamine (Figure 3.6A). Another method relies on reaction between 4-cyanopyridine and 2-vinyl pyridine followed by reaction with dimethylamine (Figure 3.6B).

3.4.2 Insoluble Polymer-supported DMAP

In the past couple of decades, there have been many variations of polymer-supported DMAP prepared and used as acylation catalysts. An early notable example is a report published by Fréchet and co-workers detailing the use of cross-linked polystyrene-supported DMAP derivatives **1–3** (Figure 3.7).[35] Active catalysts **1–3** could successfully be prepared through a suspension polymerization with styrene and divinylbenzene (as the cross-linking agent). The catalytic properties of each supported species were tested through an acylation of 1-methylcyclohexanol with acetic anhydride (Ac_2O). The authors noted that catalytic activity was dependent on "spacer length". Catalytic activity increased with spacer length ($n = 3$) as the catalyst was placed further away from the electron withdrawing benzyl moiety. Supported catalyst **3** had catalytic activity that closely resembled its low molecular weight analog **3**.

Since these early reports by Fréchet (and others) many other groups have utilized cross-linked polystyrene-supported DMAP. For instance, Hedrick and co-workers utilized this system to facilitate the ring opening polymerization of lactide, a biodegradable polyester.[33] However, because of its insolubility, this catalyst required longer reaction times than DMAP. Other

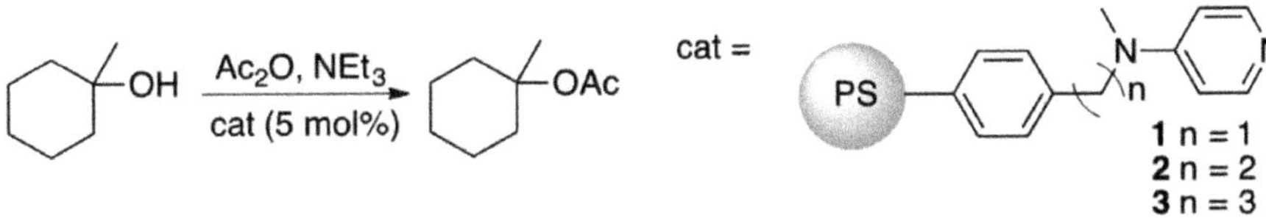

Figure 3.7 Insoluble, cross-linked PS-supported DMAPs and their use as acylation catalysts.

types of polymer-supported DMAPs have been developed and utilized as well. Toy and co-workers reported that a Rasta resin-supported DMAP showed higher catalytic activity than cross-linked PS (Merrifield resin)-supported DMAPs in the formation of cyclic carbonates by the addition of CO_2 to epoxides.[36] The higher catalytic activity could be explained by the increased solvent swelling of these flexible resins as compared to typical Merrifield-type resins. These supported catalysts could be used and recovered using typical solid/liquid separations.

An interesting report came from the laboratory of Christopher W. Jones and described the use of supported DMAPs as catalysts for regioselective epoxide ring opening (ERO) reactions with phenol.[37] Of note, they showed that a DMAP-derivative supported by magnetic nanoparticles could successfully be used to facilitate the reaction between 1,2-epoxy hexane and phenol. Advantageously, this catalyst could easily be removed from the product/toluene solution upon exposure to an external magnet. However, this catalyst showed reduced activity compared to low molecular weight DMAP and even lower activities in subsequent cycles. The authors rationalized that this was because of product adsorption onto the support.

It is often argued that insoluble polymer-supported catalysts have lower reactivity (when compared to their low molecular weight analogs) due to limited accessibility reagents have to the catalyst caused by low diffusion into the polymer environment. To avoid this issue, McQuade and co-workers developed "soluble heterogeneous catalysts" which are microencapsulated soluble polymer-supported catalysts.[38] They showed that a linear PS-supported DMAP derivative (**4**) could be encapsulated by exposure to poly(methylene-[polyphenyl]isocyanate) (PMPPI), poly(vinyl alcohol) (PVA), and tetraethylenepentamine (TEPA) (Figure 3.8). This encapsulated catalyst showed higher activity for catalyzing the reaction between *sec*-phenethyl alcohol (**5**) and AC_2O than Merrifield-supported DMAP. Furthermore, these species could easily be recovered by simple filtration and reused with no observable loss in activity. The encapsulated species still had lower activity than a soluble, linear PS-supported DMAP and low molecular weight DMAP itself.

3.4.3 Soluble Polymer-supported DMAP

The issue of lower reactivity is often a consequence of designing polymer-supported catalysts. This is especially true with regard to insoluble

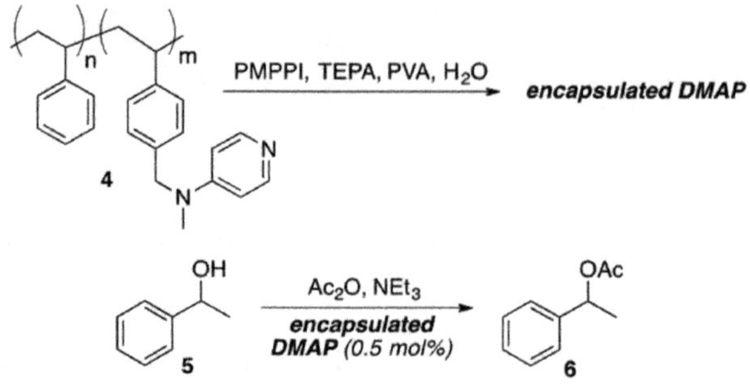

Figure 3.8 Preparation and use of encapsulated PS-supported DMAP.

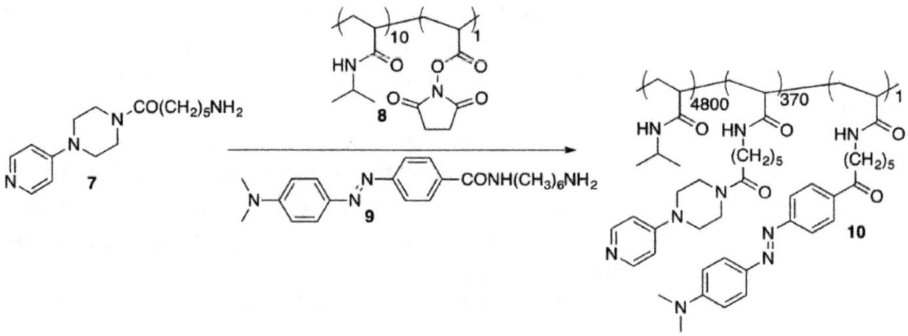

Figure 3.9 Preparation of PNIPAM-supported DMAP **10**.

polymer-supported catalysts. This can be circumvented by, instead, anchoring DMAP to a *soluble* polymer support. Although their use and recovery is less straightforward than their insoluble counterparts, they can have higher levels of reactivity. Bergbreiter's laboratory has developed many examples that utilize this idea. In doing so, they invented various ways of not only preparing these catalysts, but also recovering and reusing them. Many of these methods are described above in Section 3.3.

Their group described the use of polar, poly(*N*-isopropyl acrylamide) (PNIPAM) as a soluble catalyst support for many different types of catalysts, including DMAP.[39] Their supported DMAP-derivative (**10**) was prepared according to Figure 3.9 in which a small amount of azo dye was covalently linked to the polymer chain as well. This served as a colorimetric tag so that the authors could visually confirm polymer isolation. This PNIPAM-supported DMAP could be used as a soluble catalyst for the acylation of 1-methylcyclohexanol with Ac_2O and the Boc-protection of hindered phenols. In both cases, this supported catalyst showed lower activity than DMAP, but could be recovered and reused several times with no loss in activity; a feat that cannot be accomplished with DMAP alone. Two different recovery strategies could be used: either a precipitation in which the catalyst is isolated as a solid after the addition of hexane or a liquid–liquid separation in which the catalyst and product are separated in opposite solvent layers.

The same group later developed nonpolar analogs of PNIPAM that could be utilized under latent-biphasic separation schemes in which the polymer-supported catalyst is isolated in the nonpolar solvent phase (Figure 3.1). This is an improvement on earlier chemistry that relied on catalyst isolation in the polar phase. This is because low molecular weight products have higher solubility in slightly polar media while these nonpolar polymer-supported catalysts are highly selectively soluble for solvents such as heptane. The authors were able to prepare the nonpolar, poly(*N*-octadecylacrylamide) (PNODAM)-based supports[40] according to Figure 3.10 in which free amine-containing azo dye and DMAP-derivative (**12**) were attached through reaction with PNODAM-activated esters. This system was used to facilitate the Boc-protection of hindered phenols and could be recovered and reused up to seven times with no loss in catalytic activity. Recovery of this catalyst could be accomplished in either a monophasic mixture of heptane and ethanol (in which catalyst and product separation was achieved upon addition of 10% water) or a biphasic mixture of heptane and DMF (in which homogeneous reaction occurs at elevated temperature and catalyst/product separation is achieved upon cooling), as shown earlier in Figure 3.3. However, no comparison to low molecular weight DMAP was provided.

They later described similar chemistry utilizing linear poly(4-*tert*-butylstyrene) (PTBS) as a soluble support.[41] This system has no advantage over linear PS if catalyst recovery and reuse is facilitated using solvent precipitations. However, PTBS exhibits high phase-selective solubility in nonpolar solvents. So, PTBS (prepared according to Figure 3.11) could be used as a recoverable catalyst under both latent-biphasic and thermomorphic liquid–liquid separations (Figure 3.3) and could be reused through twenty cycles.

Using a synthetic strategy similar to that described by Bergbreiter, Toy and co-workers prepared a soluble, linear PS-supported bifunctional catalytic system in which both piperazine and a DMAP-derivative were covalently linked to linear PS (Figure 3.12).[42] This system was then used to facilitate

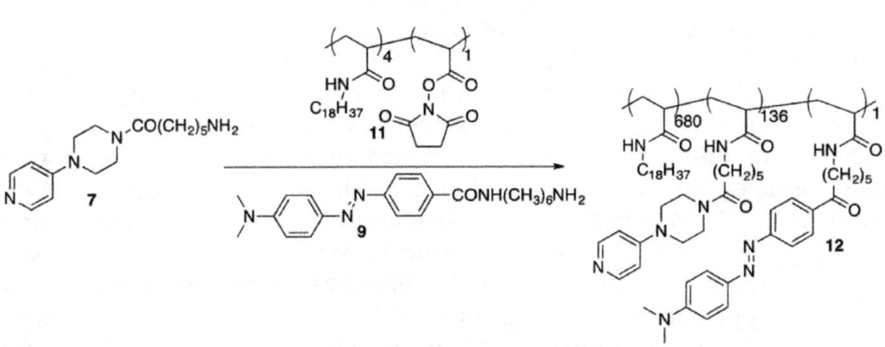

Figure 3.10 Preparation of PNODAM-supported DMAP 12.

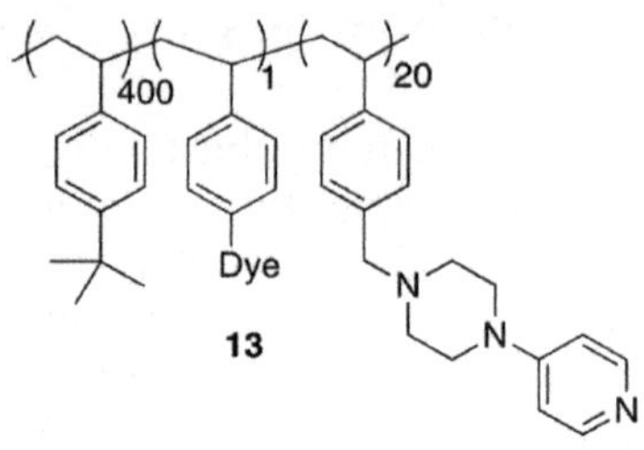

Figure 3.11 PTBS-supported DMAP.

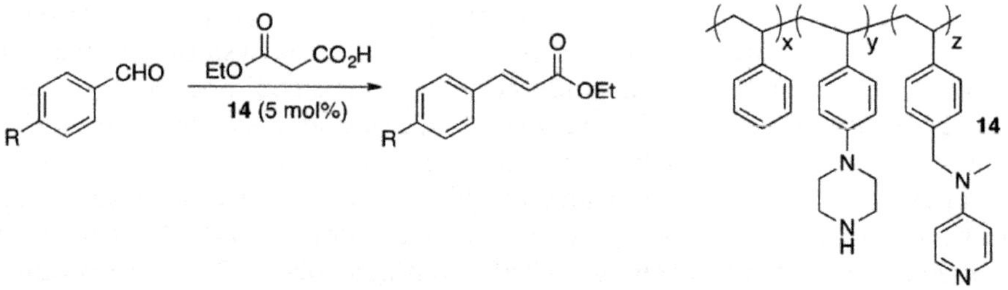

Figure 3.12 Preparation and use of bifunctional **14** for Doebner–Knoevenagel reactions.

decarboxylative, Doebner–Knoevenagel condensation reactions between various aldehydes and monoethylmalonate. However, no recycling data for this system was reported.

About this same time, Bergbreiter's laboratory reported an improvement on linear PS supports by preparing and utilizing poly(4-dodecylstyrene)[26] supports. Incorporation of the greasy dodecyl chain improved phase selectivity (over the *tert*-butyl analog) dramatically. They were able to show that this system could successfully be used and reused up to 20 times.

Shortly after this, Hobbs' laboratory reported the preparation of a terminally functionalized polymer-supported DMAP-derivative based on commercially-available isotactic-poly(propylene-co-hexene) (iPPH) oligomers.[43] This was inspired by the work of Bergbreiter's laboratory utilizing polyisobutylene (PIB) as a soluble support.[11,21,28] Furthermore, a study by this group showed that soluble, terminally functionalized polymer supports have reactivity that is virtually identical to their low molecular weight analogs.[44] With this in mind, Hobbs and co-workers prepared an iPPH-supported DMAP utilizing the click strategy shown in Figure 3.13. Subjecting olefin-terminated iPPH **15** to hydroboration/oxidation, followed by mesylation and substitution led to iPPH-terminated azide **17**. With the goals of atom economy and green chemistry in mind, a copper-catalyzed alkyne/azide click (CuAAC[45]) approach was undertaken in order to prepare the desired polymer-supported DMAP. Subjecting **17** to reaction with CuBr, PMDETA, and alkyne **19** (prepared by low temperature deprotonation followed by reaction

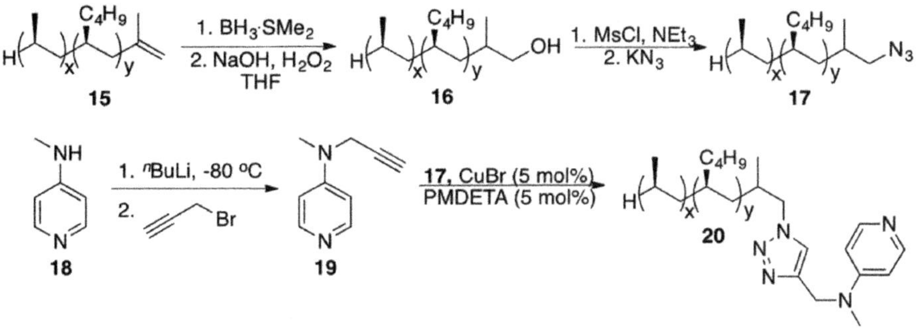

Figure 3.13 Preparation of iPPH-supported DMAP **20**.

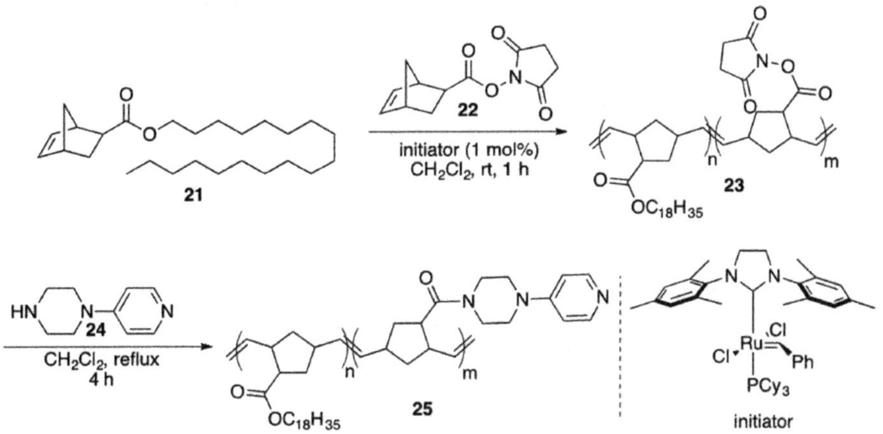

Figure 3.14 Preparation of PNB-supported DMAP **25**.

with propargyl bromide) led to iPPH-DMAP **20**. The catalytic activity of **20** was confirmed for the Boc-protection of a hindered phenol. Because iPPH exists as a semicrystalline solid, its reuse could be carried out using solvent precipitations (Figure 3.1). However, its nonpolar nature provides iPPH with high levels of phase-selective solubility. So, it could also be reused under liquid–liquid conditions as well, in which the catalyst and products are isolated in opposing solvent phases (Figure 3.3). Using either system, **20** could be recovered and recycled up to four times with no loss in yield.

Recently, Hobbs and co-workers have reported that it is also possible to design phase-selectively-soluble polynorbornenes (PNB) and utilize them as phase anchors for DMAP derivatives.[46] This approach is advantageous because it takes advantage of the well-controlled and functional group tolerant, Ring Opening Metathesis Polymerization (ROMP)[47] for the synthesis of the supporting polymer, as shown in Figure 3.14.

The preparation of this supported species relied on the copolymerization of nonpolar **21** with activated ester **22** in the presence of Grubbs 2nd generation initiator. The resulting material could then be subjected to a post-polymerization modification (through a reaction with **24**) for the installation of the catalytically-active DMAP-derivative to provide polymer-supported DMAP **25**. **25** was successfully used as a catalyst for the Boc-protection of 2,6-dimethylphenol. **25** could successfully be used up to five times and was recovered under liquid/liquid separations (Figure 3.3).

3.5 Polymer-supported DBU

In addition to DMAP, other amines have been used as competent organic bases and catalysts. For example, DBU (1,8-diazabicyclo[5.4.0]undec-7-ene) has long been used as such for a number of transformations (Figure 3.15).[48] For the same reasons described earlier in this chapter, the preparation and use of recyclable and reusable DBU derivatives is important. This has been achieved through the use of insoluble polymer supports as anchors for DBU.[49–52]

Tomoi and co-workers[49] reported an early example of the use of a cross-linked PS-supported DBU derivative that was shown to facilitate dehydrobromination and esterification reactions. The preparation of these species relied on initial deprotonation of DBU, followed by nucleophilic substitution of this anion with chloromethylated or bromoalkylated PS resins (Figure 3.16).

The authors were able to show that supported **27** could be used to facilitate the dehydrohalogenation of either 2-bromoethylbenzene or 1-bromooctane to provide styrene or 1-octene, respectively (Figure 3.17). Unfortunately, it was noted that the supported catalyst had lower activity than low molecular weight DBU. The authors concluded that this may be due to the accumulation of salts that reduce the swellability of the solid support as well as steric hindrance, resulting from the placement of the covalent linker.

They further showed that **27** could be used as a catalyst for the esterification of benzoic acid with 1-bromobutane (Figure 3.18). This supported species fared better than it did for the eliminations discussed above, as it had reactivity comparable to low molecular weight DBU.

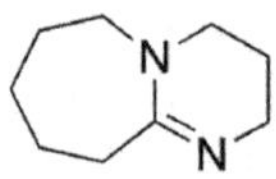

Figure 3.15 Structure of DBU.

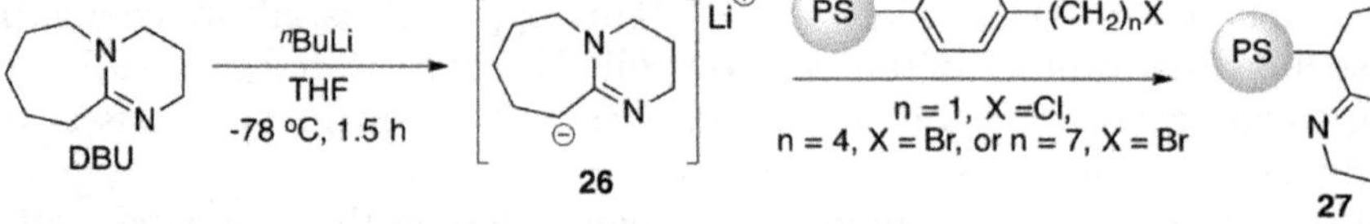

Figure 3.16 Synthesis of PS-supported DBU derivative.

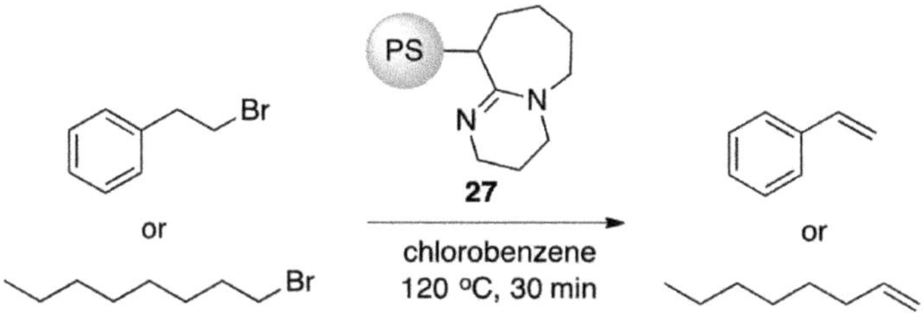

Figure 3.17 Dehydrohalogenation of 2-bromoethylbenzene or bromooctane using PS-supported DBU **27** as catalyst.

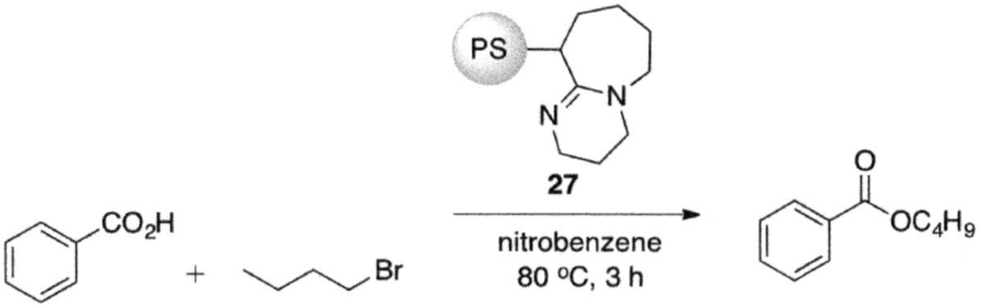

Figure 3.18 Esterification of benzoic acid using PS-supported DBU **27** as catalyst.

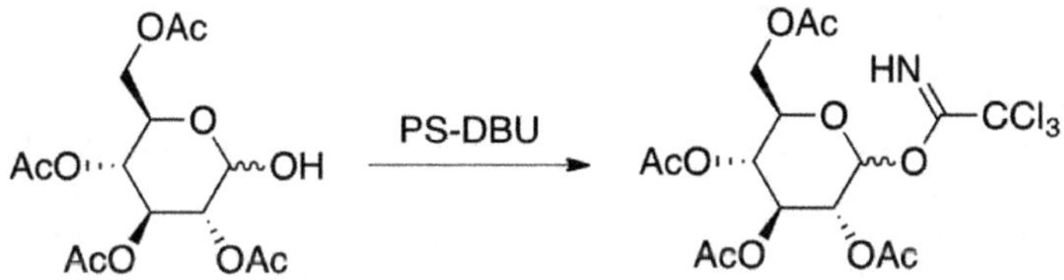

Figure 3.19 Use of PS-supported DBU as a catalyst for the synthesis of glycosyl trichloroacetimidates.

More recently, Chiara and co-workers utilized a polymer-supported DBU as a catalyst for the synthesis of glycosyl trichloroacetimidates (Figure 3.19).[53] PS-supported DBU (among other supported bases) was able to facilitate this reaction, providing products in near-quantitative yields. The PS-DBU, however, showed poor recycling; subsequent reaction yields dropped to as low as 40%.

Even more recently, Quaranta[50] *et al.* have shown that a PS-bound DBU could be used to activate a carbonic acid diester which could be used as a route towards catalytic *N*-carbonylations. It was found that diphenyl carbonate (DPC) could successfully be used as a carbonylation agent for N-heterocycles such as pyrrole, indole, and carbazole in the presence of catalytic amounts of DBU at room temperature (Figure 3.20). In this case, the authors note that DBU can act as a base, in which deprotonation activates

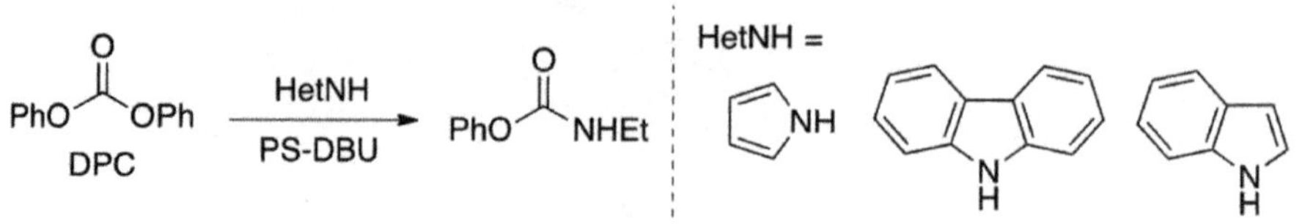

Figure 3.20 Use of PS-DBU as a catalyst for the *N*-carbonylation of N-heterocycles with DPC.

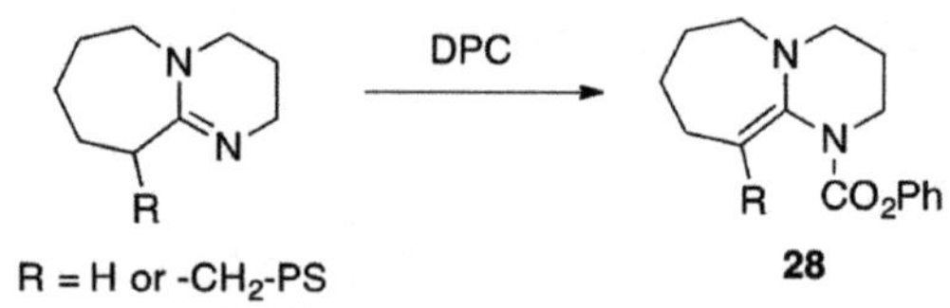

Figure 3.21 Formation of ketene aminal **28**.

the N-heterocycle, as well as a nucleophile, in which the DPC is activated through the formation of a ketene aminal **28** (Figure 3.21). PS-DBU showed only slightly less reactivity than low molecular weight DBU. The catalyst could successfully be recovered and reused up to 13 times with almost no loss in activity.

3.6 Polymer-supported TBD

In addition to DMAP and DBU, bicyclic guanidines such as 1,5,7-triazabicyclo[4.4.0]dec-5-ene (TBD) have also found widespread use in organic synthesis.[54] DBU (Figure 3.22) has been shown to catalyze Michael reactions, Henry reactions, Wittig reactions, and many more. Because of its importance, many research groups have been interested in developing and using immobilized derivatives with the goal of sequestration and reuse.

In an early example, Tomoi and co-workers showed that a TBD derivative **29**, immobilized onto an insoluble support (cross-linked PS) could serve as a useful catalyst for the esterification of benzoic acid with bromobutane (Figures 3.23 and 3.24).[55]

Schuchardt *et al.* later showed that a similar PS-supported TBD could successfully be used to catalyze the transesterification of vegetable oils with methanol,[56] an important concept with regard to green and sustainable chemistry. They reported that this immobilized TBD could be used to prepare methyl esters of soybean oil with rates only slightly lower than low molecular weight TBD, while exhibiting identical product yields.

Shortly after, Xu and co-workers reported on a combinatorial approach to the synthesis of a variety of aryl ethers utilizing a supported TBD catalyst (Figure 3.25).[57] Etherifications were carried out in acetonitrile at room temperature. Most reactions provided products in high yields (>90%) and could even be used for sterically hindered phenols.

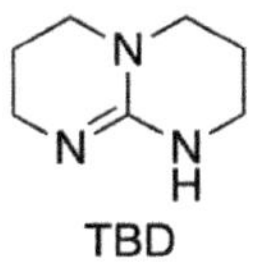

Figure 3.22 Structure of TBD.

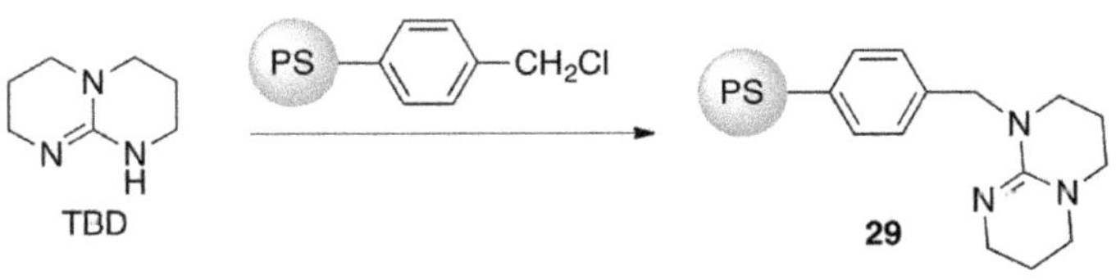

Figure 3.23 Synthesis of PS-supported TBD **29**.

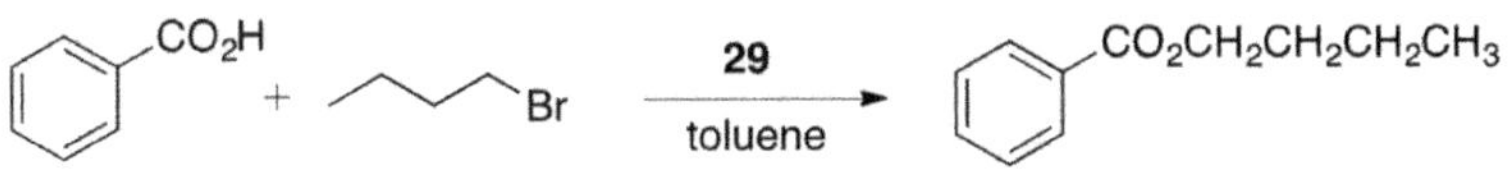

Figure 3.24 Use of PS-supported TBD **29** as an esterification catalyst.

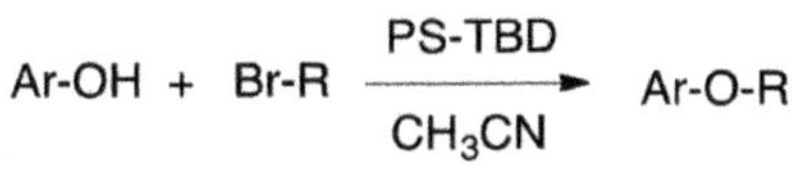

Figure 3.25 Use of PS-TBD as a catalyst for the synthesis of aryl ethers.

Simoni later showed that a PS-TBD could be used as a catalyst for the direct and conjugate addition of phosphites to carbonyls and a,b-unsaturated carbonyls, respectively as well as Henry reactions (Figure 3.26).[58] The reactions described all proceeded under mild conditions at room temperature or lower and provided products in good yields, in some cases in a matter of a few minutes.

More recently, Lanari and Vaccaro showed that a JandaJel supported TBD could successfully be used as a catalyst for Michael additions and epoxide opening reactions under solvent-free conditions.[59] The authors noted that such supported species showed catalytic activity that surpassed previously developed PS-supported catalysts and showed that these species could be recovered and reused as well with no observable loss in activity (Figure 3.27).

In 2016, Matsukawa and co-workers reported that a PS-supported TBD could be used to facilitate the cyanation of carbonyls utilizing methyl cyanoformate as a cyanation reagent (Figure 3.28).[60] This was shown to be an advantageous route as the cyanoformates are more stable and less prone to hydrolysis when compared to trimethylsilyl cyanide (TMSCN). The authors showed that the supported catalyst showed only slightly slowed activity (when compared to low molecular weight TBD) in reactions with aldehydes

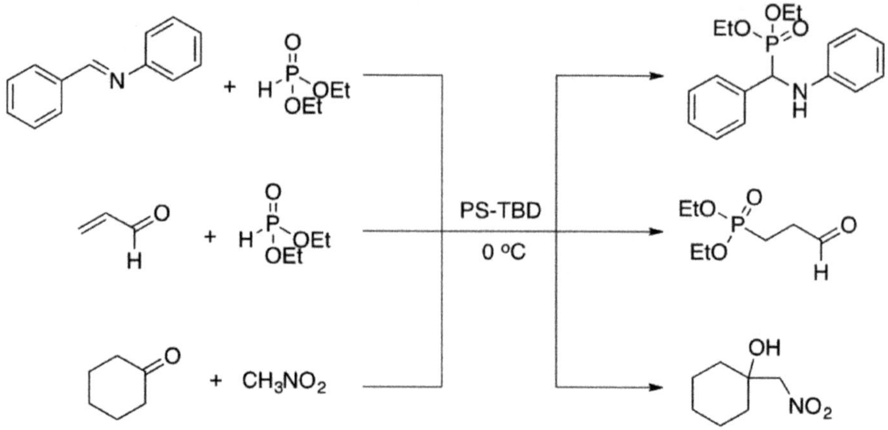

Figure 3.26 Some examples of the use of polymer-supported TBD as a catalyst for carbonyl additions.

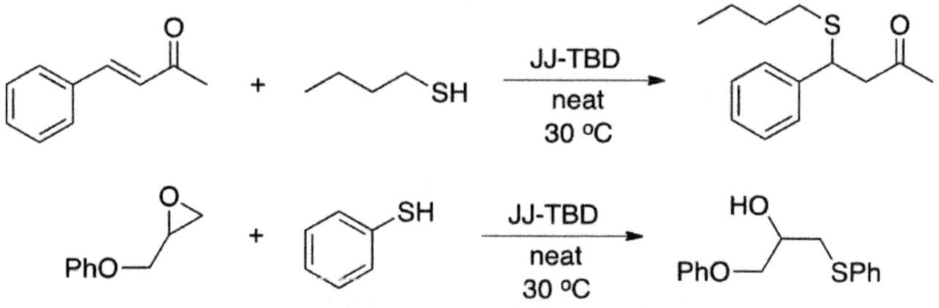

Figure 3.27 Some examples of JandJel-TBD facilitated reactions.

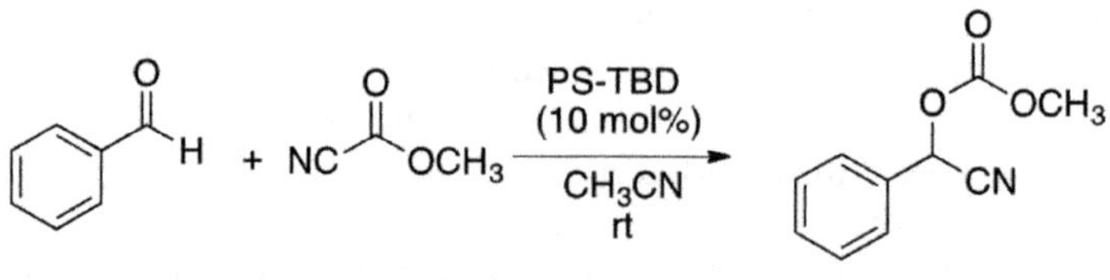

Figure 3.28 Some examples of PS-TBD facilitated cyanation reactions.

but required elevated temperatures for reactions with ketones. But, the supported catalyst could be recovered and reused with no loss in activity three times.

3.7 Conclusion

As the scientific community continues to become more cognizant of important environmental factors, the field of polymer-supported catalysis will continue to grow as this allows chemists to recycle important species as well as reduce the overall carbon footprint of synthesis. This is especially important in the face of global climate change, in which the utilization of the principles of Green Chemistry will only become more and more vital, not only to chemistry, but to the general population. Although this chapter focuses on supported N-heterocyclic organocatalysts, there exist many other elegant examples of underlying the importance that this field will continue to have in synthesis.

References

1. *Handbook of Green Chemistry: Green Catalysis*; R. H. Crabtree, P. T. Anastas, Wiley-VCH: Weinheim, Germany, 1st edn, 2009.
2. X. Bugaut and F. Glorius, *Chem. Soc. Rev.*, 2012, **41**, 3511.
3. R. A. Sheldon, *Chem. Soc. Rev.*, 2012, **41**, 1437.
4. D. Enders, O. Niemeier and A. Henseler, *Chem. Rev.*, 2007, **107**, 5606.
5. D. J. Darensbourg, *Chem. Rev.*, 2007, **107**, 2388.
6. *Handbook of Fluorous C*hemistry, J. A. Gladysz, D. P. Curran, I. T. Horvath, Wiley-VCH: Weinheim, Germany, 1st edn, 2004.
7. *Ionic Liquids in Synthesis*, P. Wassersheid, T. Welton, Wiley-VCH: Weinheim, Germany, 2nd edn, 2008.
8. P. G. Jessop, *J. Supercrit. Fluids*, 2006, **38**, 211.
9. A. Nachtergael, O. Coulembler, P. Dubois, M. Helvenstein, P. Duez, B. Blankert and L. Mespouille, *Biomacromolecules*, 2015, **16**, 507.
10. *Recoverable and Recyclable Catalysts*, M. Benaglia, Wiley-VCH: Weinheim, Germany, 1st edn, 2009.
11. D. E. Bergbreiter, *Chem. Rev.*, 2002, **102**, 3345.
12. D. E. Bergbreiter, J. Tian and C. Hongfa, *Chem. Rev.*, 2009, **109**, 530.
13. J. Lu and P. H. Toy, *Chem. Rev.*, 2009, **109**, 815.
14. C. S. Slater and M. A. Savelski, *J. Environ. Sci. Health Part A*, 2007, **42**, 1595.
15. R. B. Merrifield, *J. Am. Chem. Soc.*, 1963, **85**, 2149.
16. R. L. Letsinger and M. J. Kornet, *J. Am. Chem. Soc.*, 1963, **85**, 3045.
17. D. D. Whitehurst, *CHEMTECH*, 1980, **10**, 44.
18. E. Bayer and V. Schurig, *Angew Chem., Int. Ed. Engl.*, 1975, **14**, 493.
19. E. Bayer, H. Eberhardt and K. Geckeler, *Angew. Makromol. Chem.*, 1981, **97**, 217.
20. D. E. Bergbreiter and R. Chandran, *J. Am. Chem. Soc.*, 1987, **109**, 174.
21. D. E. Bergbreiter, C. Hobbs and C. Hongfa, *J. Org. Chem.*, 2011, **76**, 523.
22. C. Hobbs, Y.-C. Yang, J. Ling, S. Nicola, H.-L. Su, H. S. Bazzi and D. E. Bergbreiter, *Org. Lett.*, 2011, **13**, 3904.
23. Y. Yang, N. Priyadarshani, T. Khamatnurova, J. Suriboot and D. E. Bergbreiter, *J. Am. Chem. Soc.*, 2012, **134**, 14714.
24. J. Suriboot, C. E. Hobbs, W. Guzman, H. S. Bazzi and D. E. Bergbreiter, *Macromolecules*, 2015, **48**, 5511.
25. C. M. Older, S. Kristjandsdottir, J. C. Ritter, W. Tam and M. C. Grady, *Chem. Ind.*, 2009, **123**, 319.
26. T. Khamatnurova and D. E. Bergbreiter, *Polym. Chem.*, 2013, **4**, 1617.

27. D. E. Bergbreiter, N. A. Avilés-Ramos and D. Ortiz-Acosta, *J. Comb. Chem.*, 2007, **9**, 609.
28. C. Hongfa, H.-L. Su, H. S. Bazzi and D. E. Bergbreiter, *Org. Lett.*, 2009, **11**, 665.
29. B. Ling, D. Lawler, G. P. McGovern, C. A. Bradley and C. E. Hobbs, *Tetrahedron Lett.*, 2013, **54**, 970.
30. C. M. R. Volla, I. Atodiresei and M. Rueping, *Chem. Rev.*, 2014, **114**, 2390.
31. G. Hoefle, W. Steglich and H. Vorbrueggen, *Angew. Chem. Int. Ed.*, 1978, **17**, 569.
32. E. F. Scriven, *Chem. Soc. Rev.*, 1983, **12**, 129.
33. F. Nederberg, E. F. Conner, M. Möller, T. Glauser and J. L. Hedrick, *Angew. Chem. Int. Ed.*, 2001, **40**, 2712.
34. S. Shimizu, N. Watanabe, T. Kataoka, T. Shoji, N. Abe, S. Morishita and H. Ichimura, *Ullman's En. Ind. Chem.*, 2000, 559–564.
35. A. Deratani, G. D. Darling, D. Horak and J. M. J. Fréchet, *Macromolecules*, 1987, **20**, 767.
36. J. Lu and P. H. Toy, *Synlett*, 2011, **5**, 669.
37. N. A. Brunelli, W. Long, K. Venkatasubbaiah and C. W. Jones, *Top. Catal.*, 2012, **55**, 432.
38. K. E. Price, B. P. Mason, A. R. Bogdan, S. J. Broadwater, J. L. Steinbacher and D. T. McQuade, *J. Am. Chem. Soc.*, 2006, **128**, 10376.
39. D. E. Bergbreiter, P. L. Osburn and C. Li, *Org. Lett.*, 2002, **4**, 737.
40. D. E. Bergbreiter, P. L. Osburn, T. Smith, C. Li and J. D. Frels, *J. Am. Chem. Soc.*, 2003, **125**, 6254.
41. D. E. Bergbreiter and C. Li, *Org. Lett.*, 2003, **5**, 2445.
42. J. Lu and P. H. Toy, *Synlett*, 2011, **12**, 1723.
43. A. Hicks, B. Lin, P. L. Osburn and C. E. Hobbs, *J. Polym. Sci., Part A: Polym. Chem.*, 2013, **52**, 600.
44. D. E. Bergbreiter and Y.-C. Yang, *J. Org. Chem.*, 2010, **75**, 873.
45. V. V. Rostovtsev, L. G. Green, V. V. Fokin and K. B. Sharpless, *Angew. Chem. Int. Ed.*, 2002, **41**, 2596.
46. A. Fredlund, V. A. Kothapalli and C. E. Hobbs, *Polym. Chem.*, 2017, **8**, 516.
47. G. C. Vougioukalakis and R. H. Grubbs, *Chem. Rev.*, 2010, **110**, 1746.
48. B. Nand, G. Khanna, A. Chaudhary, A. Lumb and J. M. Khurana, *Curr. Org. Chem.*, 2015, **19**, 790.
49. M. Tomoi, Y. Kato and H. Kakiuchi, *H. Die Makromol. Chem.*, 1984, **185**, 2117.
50. E. Quaranta, A. Angelini, M. Carafa, A. Dibenedetto and V. Mele, *ACS Catal.*, 2014, **4**, 195.
51. J. L. Chiara, L. Encinas and B. Díaz, *Tetrahedron Lett.*, 2005, **46**, 2445.
52. Y. Tamura, W. Fukuda, M. Tomoi and S. Tokuyama, *Synth. Commun.*, 1994, **24**, 2907.
53. J. L. Chiara, L. Encinas and B. Díaz, *Tetrahedron Lett.*, 2005, **46**, 2445.
54. A. Huczynski, B. Brzezinski, *1,5,7-triazabicyclo[4.4.0]dec-5-ene, e-EROS*, 2008, John Wiley and Sons.

55. K. Iijima, W. Fukuda and M. Tomoi, *J. Macromol. Sci., Part A: Pure Appl. Chem.*, 1992, **29**, 249.
56. U. Schuchardt, R. M. Vargas and G. Gelbard, *J. Mol. Catal. A: Chem.*, 1996, **109**, 37.
57. W. Xu, R. Mohan and M. M. Morrissey, *Tetrahedron Lett.*, 1997, **38**, 7337.
58. D. Simoni, R. Rondanin, M. Morini, R. Baruchello and F. P. Invidiata, *Tetrahedron Lett.*, 2000, **41**, 1607.
59. D. Lanari, R. Ballini, S. Bonollo, A. Palmieri, F. Pizzo and L. Vaccaro, *Green Chem.*, 2011, **13**, 3181.
60. S. Matsukawa, J. Kimura and M. Yoshioka, *Molecules*, 2016, **21**, 1030.

CHAPTER 4

Synthesis of Atorvastatin

MARTIN EDWARD FOX

Chirotech Technology Centre, Dr Reddy's Laboratories(EU) Limited, Cambridge Science Park, Milton Road, Cambridge CB4 0PE, UK
Email: mfox@drreddys.com

4.1 Introduction

Atorvastatin **1**, marketed by Pfizer as Lipitor® and launched in 1996, is probably the most commercially successful small molecule drug ever launched, with annual sales peaking at US$12.8Bn in 2006 prior to patent expiry in 2009. It is used for treatment of hyperlipidemia and hypercholesterolemia. Together with related drugs such as pitavastatin **2** and rosuvastatin **3**, atorvastatin **1** is a fully synthetic analogue of the polyketide fungal metabolites mevastatin **4** and lovastatin **5**, also a launched drug, and derived semisynthetic drugs simvastatin **6** and pravastatin **7** (Figure 4.1).[1] These compounds are inhibitors of 3-hydroxy-3-methylglutaryl-coenzyme A (HMG-CoA) reductase, a key enzyme in the isoprenoid biosynthetic pathway from mevalonate to cholesterol.

Structurally, atorvastatin is formed from two contrasting moieties; together with other members of the class, it possesses firstly an achiral heterocyclic core unit and secondly a (3*S*, 5*S*)-3,5-dihydroxyheptanoyl side-chain identical to the acetogenic side-chain of the natural products on which it is based. Hence, its synthesis features both heterocyclic chemistry and asymmetric synthesis. It was inevitable that such a commercially important and structurally interesting class of compounds would receive large amount of attention from synthetic chemists, and this has inspired the development of much synthetic methodology. A large number of syntheses of atorvastatin

Green Chemistry Series No. 52
Hazardous Reagent Substitution: A Pharmaceutical Perspective
Edited by Rakesh Kumar Sharma and Rakeshwar Bandichhor

Published by the Royal Society of Chemistry, www.rsc.org

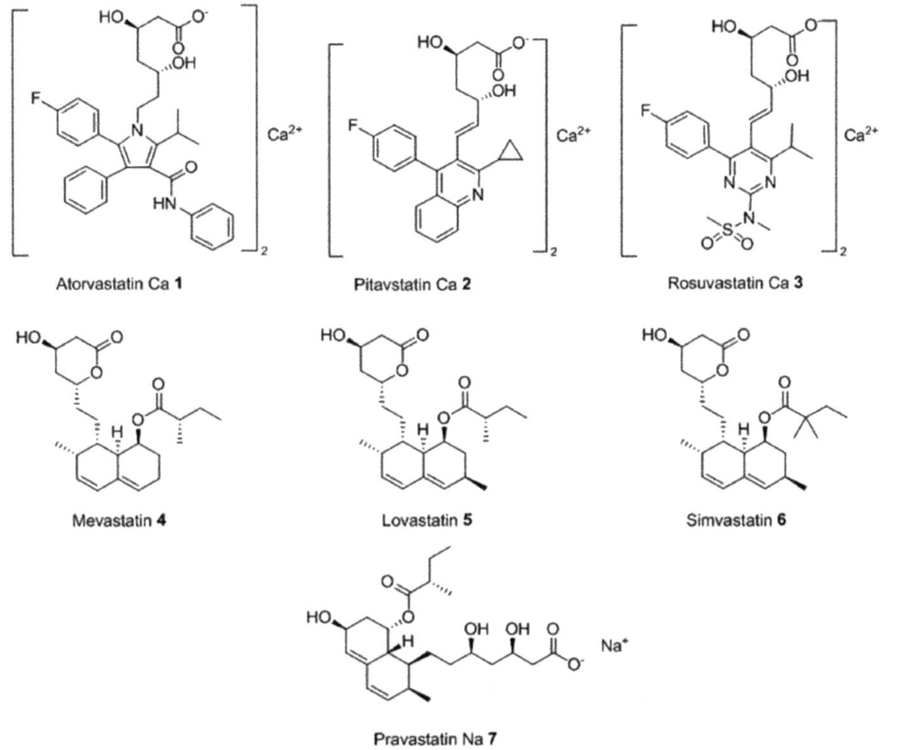

Figure 4.1 Atorvastatin and related compounds of interest.

and its synthetic intermediates have been reported. The discovery and development of atorvastatin[2,3] and approaches to its manufacture[4] have been reviewed. This overview traces the evolution of synthetic philosophy, and the inextricably linked and differing requirements of reagents through the stages of product development.

4.2 Medical Chemistry Route

4.2.1 Synthetic Strategy

In the discovery of atorvastatin at Warner–Lambert, the most important requirement of the synthetic route was the ability to vary readily substituents at positions 3 and 4 of the pyrrole ring in order for the structure–activity relationship to be established.[5] The lactone form **13** was the focus of the medicinal chemistry studies; only later was the open-chain calcium salt **1** determined as the preferred form for oral formulation. The key features of the route are flexibility, to allow the introduction of a variety of substituents, and reliability, while ensuring a minimum number of steps from a common intermediate incorporating the 2- and 5- substituents, amido-acid **9** (Schemes 4.1 and 4.2). The side-chain was introduced in truncated form,

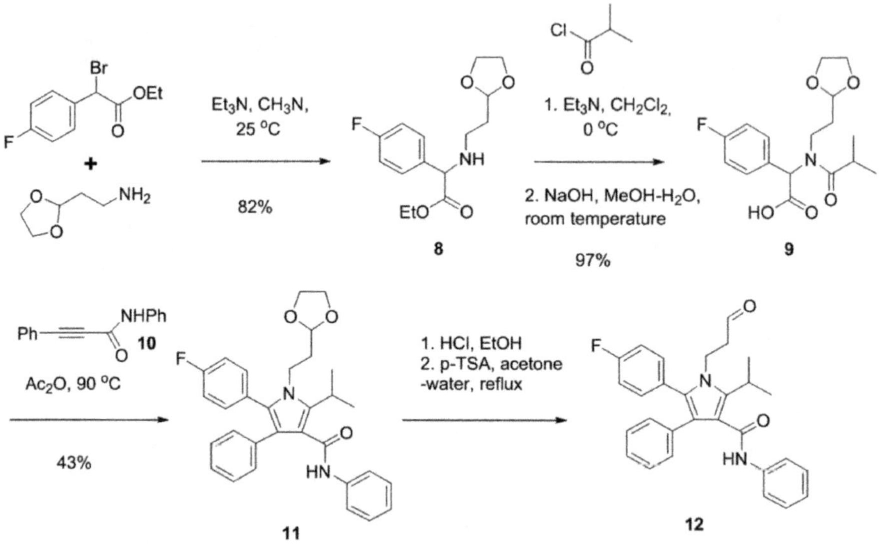

Scheme 4.1 Medicinal chemistry route to atorvastatin – pyrrole assembly.

starting with the amino-acetal **8.** The robust acetal functionality could easily be carried through the pyrrole syntheses, allowing a range of analogues to be prepared efficiently and then elaborated to the full side-chain in the final stages.

4.2.2 Synthesis of Pyrrole Unit

The method of construction of the pyrrole ring was 1,3-dipolar 3 + 2 cycloaddition of a münchnone with an electron-deficient acetylene. In this approach, the münchnone starting material delivered the groups that upon subsequent pyrrole formation became incorporated as substituents at the 2 and 5 positions, while the acetylene delivered the groups at the 3 and 4 positions of the pyrrole. Thus, when using this route, being convergent and requiring simple starting materials, the effect of variation of these substituents could be probed readily. While subsequently set aside, this approach is nevertheless concise, and may avoid scale-up issues manifested in the development and manufacturing routes. This route was re-visited some 25 years later in a more convergent implementation using the complete side-chain.[6]

The 1,3-dipole precursor, amido-acid **9** was prepared by *N*-alkylation of the ethylene glycol acetal of 3-aminopropanal with ethyl 2-bromo-2-(4-fluorophenyl)acetate (Scheme 4.1). The amine **6** was acylated with isobutyryl chloride then the ethyl ester was cleaved with sodium hydroxide. Subsequent dehydration of the acid **9** to the corresponding münchnone with acetic anhydride and cycloaddition with acetylenic amide **10** provided the pyrrole **11.** The 1,3-dioxolane was cleaved using a 2-step acid deprotection to yield aldehyde **12.**

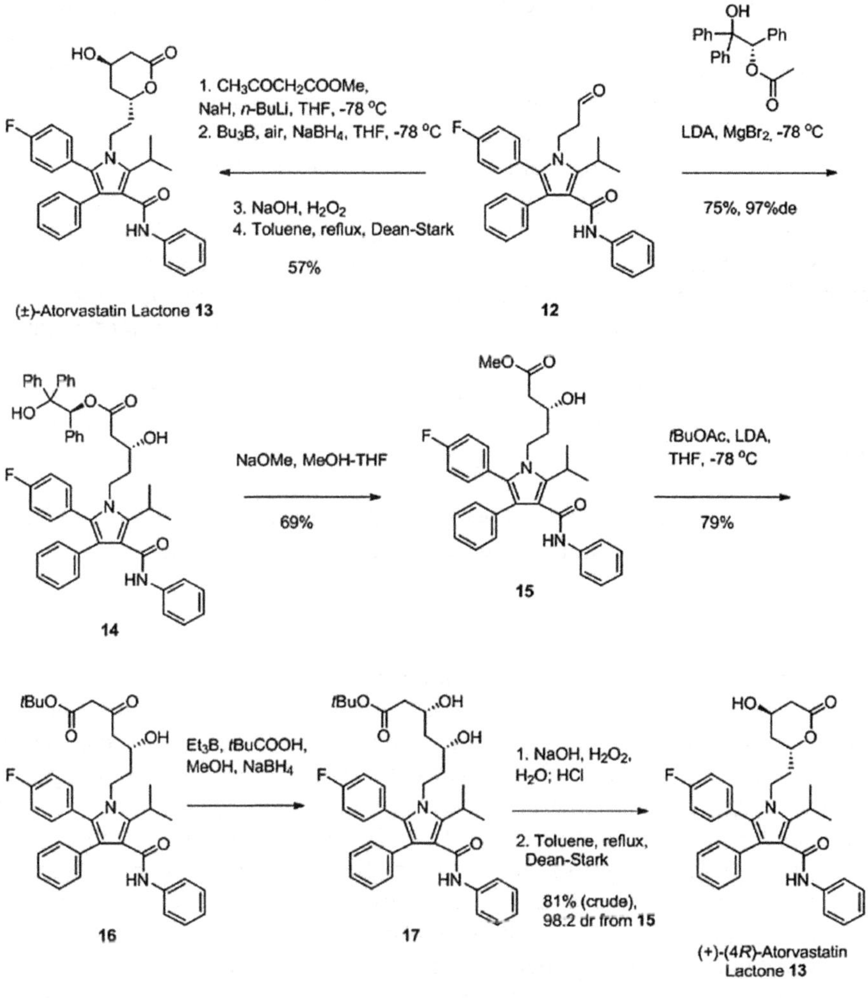

Scheme 4.2 Medicinal chemistry route to atorvastatin – side-chain elaboration.

4.2.3 Elaboration of Aldehyde Intermediate 12 to Atorvastatin

Chain extension of aldehyde **12** to atorvastatin lactone **15** involves introduction of the two stereocentres (Scheme 4.2). The methodology, especially in the stereogenic steps, possesses the primary attributes of predictability and generality rather than ease of scale-up, low hazard or potential for achieving low process-mass intensity.

For the racemate, a more convergent approach was possible. Thus, Weiler dianion condensation of aldehyde **12** with methyl acetoacetate, followed by stereoeselective reduction of the hydroxy-ketone intermediate to the *syn*-1,3-diol using the Prasad method employing tributylborane and sodium borohydride[7] base hydrolysis with sodium hydroperoxide and lactonisation provided racemic atorvastatin lactone **11**. Subsequently, asymmetric versions of the Weiler dianion condensation have been developed,[8,9] but at the time, only racemic versions were known, so synthesis of the (+)-(4*R*)-enantiomer utilized two sequential two-carbon chain-extensions. Firstly, a Braun asymmetric acetate aldol reaction employing the (*S*)-2,3,4-triphenylethane-1,2-diol auxiliary gave hydroxyl-ester **14**. After cleavage of the auxiliary by *trans*-esterification with methanol, the second 2-carbon chain extension employed a mixed Claisen condensation with *tert*-butyl acetate followed by stereoselective ketone reduction using the Prasad method, after which cleavage of the *tert*-butyl ester with sodium hydroperoxide and lactonisation completed the route.

Later development and manufacturing routes to side-chain assembly parallel the basic approach of the medicinal chemistry route in steps which mimic the action of a polyketide synthase:

1. Starting with a 5-functionalised (*S*)-3-hydroxyvalerate analogue of **15**
2. A 2-carbon chain extension with an acetate anion equivalent
3. Stereoselective ketone reduction
4. Deprotection

The Braun asymmetric acetate aldol reaction requires a high molecular weight (Mw 290) auxiliary, which while recyclable, exceeds the threshold of 220 for a reactant to be classed as a 'complex intermediate' for estimation of process-mass intensity.[10,11] Additionally, this step requires a pyrophoric organolithium reagent for generation of LDA and cryogenic reaction temperature. Replacement of this auxiliary by other sources of chirality in preparation of a 5-functionalized (*S*)-3-hydroxyvalerate precursor is a key feature of development and manufacturing routes to atorvastatin. The Claisen condensation step, which requires a near three-fold excess of lithium amide base is retained in later routes. The stereoselective ketone reduction step employs a pyrophoric organoborane reagent and energetic borohydride reductant. Substitution of this reagent combination with less hazardous alternatives has also been the focus of much development.

4.3 Process Development Route

4.3.1 Synthetic Strategy

The linear nature of the medicinal chemistry route and hence low overall yield is its principal drawback as a development route. This was addressed by adopting a more convergent strategy employing a Paal–Knorr pyrrole synthesis (Scheme 4.3).[12] The Paal–Knorr pyrrole formation employed two components, diketone **18** and side-chain amine **19** in which the 1,3-diol functionality is protected as an acetonide and the acid as a *tert*-butyl ester. The preferred form of atorvastatin for oral formulation was identified as the

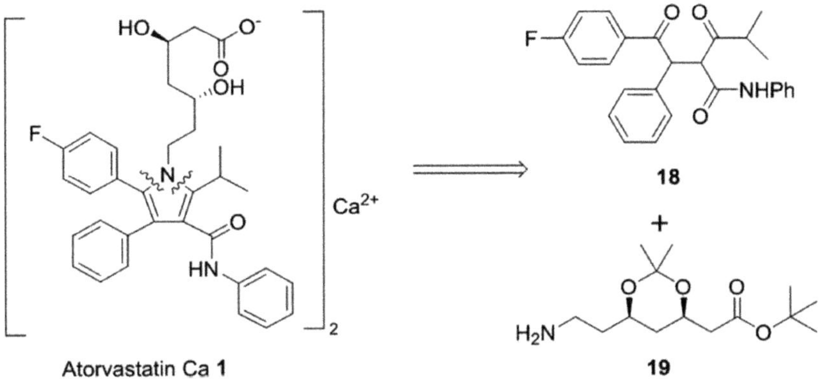

Scheme 4.3 Paal–Knorr disconnection of atorvastatin.

calcium salt of the open-chain acid,[13–15] which allowed the final deprotection and isolation steps to be defined.

4.3.2 Synthesis of 1,4-Diketone Component 18

As a 1,4-dicarbonyl compound, diketone **18** requires an *umpolung* strategy. The route started with preparation of isobutyryl acetanilide **20** from methyl isobutyrlacetate and aniline in the presence of 10% ethylenediamine (Scheme 4.4). Subsequently, Knoevenagel condensation with benzaldehyde and Stetter reaction of the benzylidene adduct **21** with 4-fluorobenzaldehyde using thiazolium salt **22** as a catalyst provided diketone **18**. All three steps employ a non-hazardous organocatalyst, albeit at high loadings. The early part of the medicinal chemistry route in which the pyrrole core is assembled uses a benzylic 2-bromo-ester, a likely genotoxic compound, and also a volatile (bp 92 °C) and reactive acid chloride. Therefore, the use of the Paal–Knorr pyrrole synthesis removes the need for both of these reactants.

4.3.3 Synthesis of Amine Side-chain Component 19

The side-chain amine component **19** is a small and relatively densely functionalised molecule possessing two stereocentres. Development of an efficient route to this component was a significant challenge in the process development of atorvastatin, and several approaches were investigated.[16,17] Synthesis of the side-chain of the statins has been the subject of much subsequent research and many innovative routes have been developed.[18–20]

Routes used in process development followed the same general approach as the latter part of the medicinal chemistry route to the side-chain, as discussed in Section 4.2.3. However, instead of the complete pyrrole moiety, they incorporate functionality suitable for later introduction to the amino group by cyanide displacement and reduction (Scheme 4.5). Starting with a 5-carbon precursor containing the 4(R) stereocentre, introduction of the

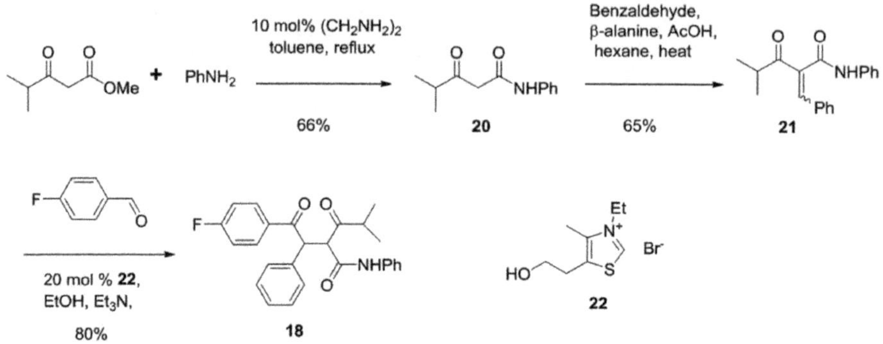

Scheme 4.4 Process development route – synthesis of diketone **18**.

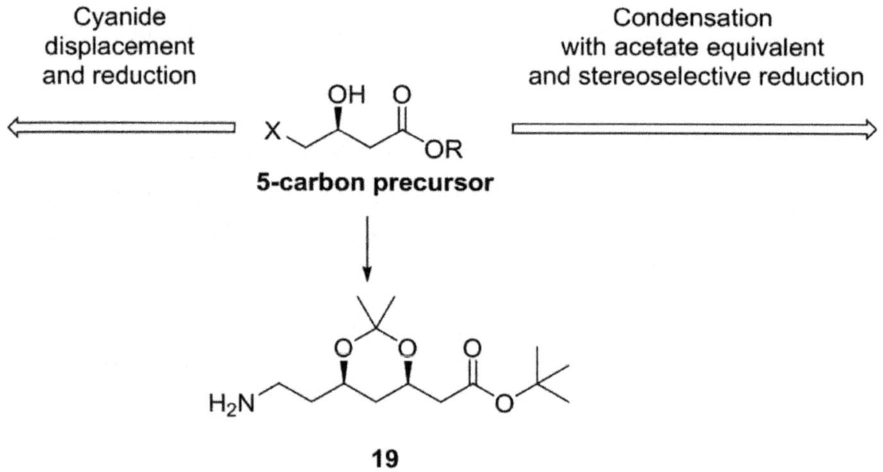

Scheme 4.5 Strategy of development approaches to side-chain amine **19**.

cyanide group could be carried out either before or after the 2-carbon homologation/stereoselective reduction sequence at the opposite chain terminus.

An expedient route used in development employed alcohol **28**, commercially available at the time, as an intermediate (Scheme 4.7). In this route, the cyanation step was carried out at a late stage. A route to alcohol **28** using L-malic acid as the 5-carbon starting material was developed by Kaneka[21] following the general approach described above. Starting with the acetonide of L-malic acid (Scheme 4.6), the free acid group was activated with carbonyldiimidazole, and condensation with magnesium bis(mono-*tert*-butyl) malonate was used for the 2-carbon homologation. This method avoids the use of LDA required for generation of acetate enolates. After methanolysis of the acetonide group, the hydroxy-diester **26** was reduced stereoselectively using the alkylborane mediated reduction. Finally, acetonide protection of the diol and selective reduction of the methyl ester

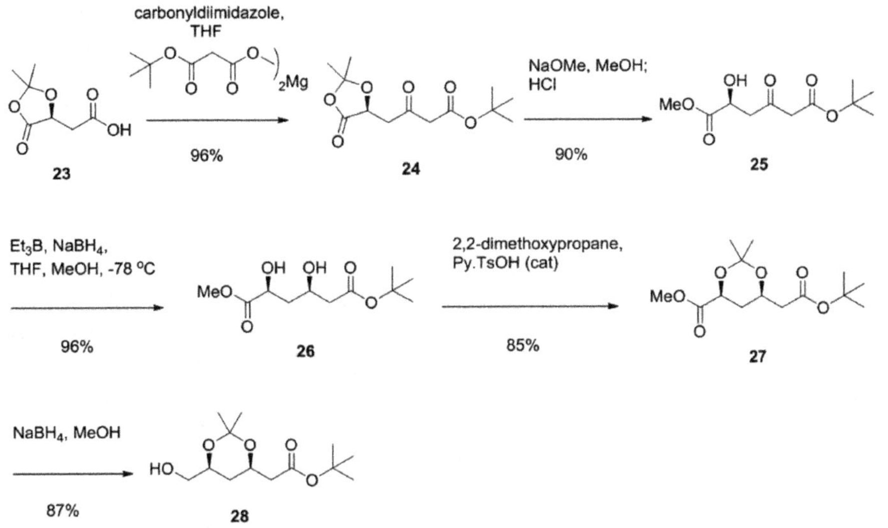

Scheme 4.6 Synthesis of side-chain alcohol **28** from malic acid.

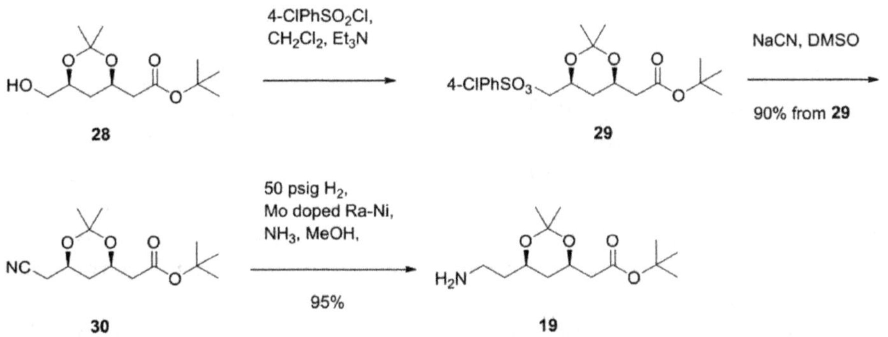

Scheme 4.7 Synthesis of side-chain amine **19** from alcohol **28**.

provided the key alcohol intermediate **28**. Earlier routes to this intermediate, including a less efficient route starting with from L-malic acid,[22] and a route starting with (*R*)-epichlorohydrin[23] also used 2-carbon homologation and stereoselective reduction steps as outlined above. A route using asymmetric ketone hydrogenation instead of L-malic acid as a source of chirality but otherwise using similar steps has also been described.[24] Subsequently, routes to this intermediate using less hazardous and more highly stereoselective biocatalytic methods for the stereoselective reduction step have been developed.[25,26]

The activation of the primary hydroxyl group of alcohol **28** as the 4-chlorobenzenesulfonate ester and substitution with cyanide, while high-yielding, posed significant challenges. Thus, the sulfonate **31** was sensitive to traces of moisture or acid, and the cyanide displacement reaction required precise control over stoichiometry and reaction conditions. Significantly, the nitrile **30** was found to be a crystalline solid, providing a valuable purification point. The final step, hydrogenation of the nitrile **30** using a Raney nickel catalyst in the presence of ammonia was clean and high-yielding. Owing to these benefits, combined with the ease of purification of nitrile **30**, this step has been retained in nearly all later routes to amine **19**.

In a second approach used in process development the cyanide displacement was placed before the 2-cartbon homologation and stereoselective reduction steps. In this approach, the 5-carbon precursor was the bromo-ester **31** (Scheme 4.8).[16,17] A modification of this route using protection of the alcohol as a *tert*-butyldimethylsilyl (TBDMS) group was used originally, but is longer and less efficient. The modest yield in the cyanation of bromo-ester **33** reflects the base-instability of the epoxide intermediate in this reaction. From the nitrile-alcohol **34**, the mixed Claisen acetate condensation and stereoselective ketone reduction sequence used in the medicinal chemistry route was employed. The monomalonatate chemistry, which avoids the use of strong amide bases such as LDA, cannot be used in the presence of the free hydroxyl group and its use requires this to be protected.[17] In the stereoselective ketone reduction step, commercial diethylmethoxyborane was used, avoiding the hazardous *in situ* generation from triethylborane. After protection of diol **34**, the route converges onto the route of Scheme 4.7 at the acetonide **30**.

The main drawback of the route from bromo-alcohol **31** as a development route was the relatively inefficient chiral pool based chemistry available for synthesis of this material (Scheme 4.9)[27–31] Starting with isoascorbic acid **35**, oxidative cleavage of the double bond with basic hydrogen peroxide provided potassium D-erythronate **36**. Bromination with hydrobromic acid in methanol was selective for the the primary hydroxyl and hydroxyl group alpha to the ester. Selective hydrogenolysis of the 2-bromide gave the methyl bromo-ester **38**.

A second route to cyano-ester **32** employed L-malic acid as the chiral starting material (Scheme 4.10).[16] In an alternative approach to selective

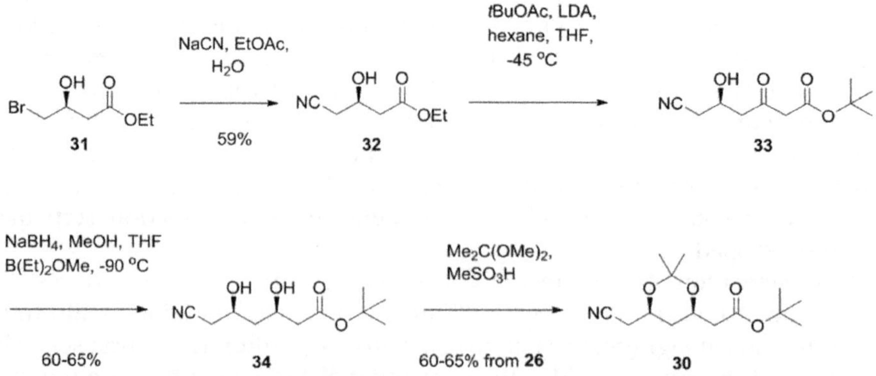

Scheme 4.8 Synthesis of amine **19** – early introduction of nitrile group.

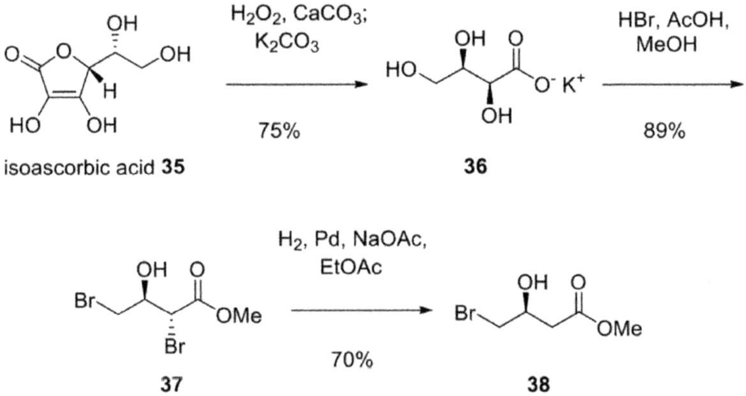

Scheme 4.9 Synthesis of bromo-ester **38** en route to amine **19**.

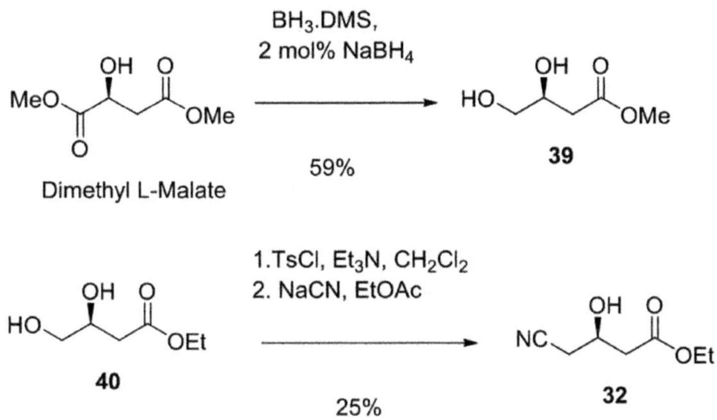

Scheme 4.10 Synthesis of cyano-ester **32** from malic acid.

modification of the carboxylate groups of malic acid to that shown in Scheme 4.6, dimethyl L-malate was be reduced selectively with borane dimethylsulfide in the presence of catalytic sodium borohydride.[27] In addition to the disadvantage of requiring hazardous borane for this reduction, selective tosylation followed by cyanide displacement was low-yielding and gave the product **26** in low purity.

Subsequently, Hollingsworth[32] has described a method of conversion of diol **39** to the bromo-ester **31**, (Scheme 4.8) *via* the delta-lactone **41** (Scheme 4.11).

Additionally, methods of preparation of lactone **41** by oxidative degradation of inexpensive carbohydrate precursors such as lactose and maltose with basic hydrogen peroxide were found.[33–35] However, despite together providing overall a more efficient entry to the bromo-ester **31**, the approach combining these methods still retains the need for a hazardous oxidant (basic hydrogen peroxide) and hydrobromic acid.

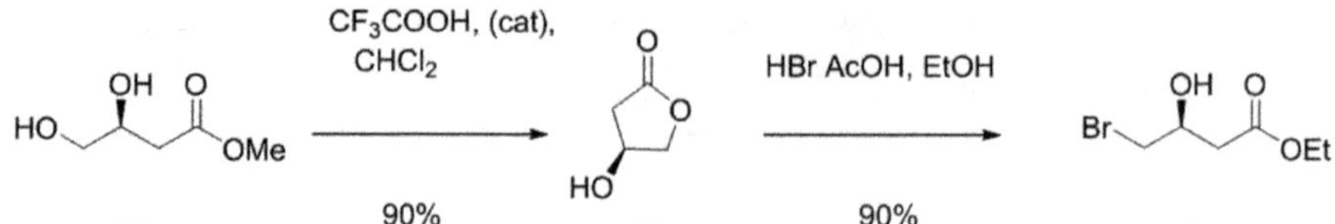

39 41 31

Scheme 4.11 Synthesis of bromo-ester **38** from diol precursor *via* delta-lactone **41**.

In development, routes to the side-chain amine **19** suitable for scale-up were discovered, but nevertheless significant remaining drawbacks were found. The use of chiral pool starting materials accounts for several hazardous reagents; functional manipulation of these requires either borane or borohydride as a reductant or hydrogen peroxide and hydrobromic acid. All routes relied on alkylboron-based reductants in the key stereoselective ketone reduction step. Development of more efficient and less hazardous routes to this component is a feature of subsequent evolution of the chemistry towards atorvastatin. Use of cyanide-based chemistry for introduction of the nitrogen atom was common to all process development routes. While concise and high-yielding routes which avoid cyanide have subsequently been developed, for example an aldolase-based route,[36] problems remain. For example, in this route, extra functional manipulation steps are required to install the desirable *tert*-butyl ester. The final Raney nickel-based hydrogenation of the nitrile functionality is another common step, and while alternative, non-pyrophoric hydrogenation catalysts able to operate in the absence of ammonia are known,[37] their cost-effectiveness is so far insufficient for them to have been widely deployed industrially.

4.3.4 Paal–Knorr Reaction and Downstream Steps

Owing to the hindered nature of diketone **18**, the condensation with amine **19** (Scheme 4.12) is slow, and extensive optimization of the reaction was required, and ultimately the use of pivalic acid as a catalyst in refluxing toluene–heptane–THF was found to clean and high-yielding. From di-protected atorvastatin **42**, deprotection and calcium salt formation were achieved without isolation of intermediates. The acetonide group was removed with HCl in methanol, converging with the medicinal chemistry route at diol **17** (Scheme 4.12). The *tert*-butyl ester was cleaved with sodium hydroxide; anchimeric assistance from the 6-hydroxyl group making deprotection of this hindered ester relatively straightforward. Hence, the hazardous peroxide-based deprotection used in the medicinal chemistry route was eliminated. The calcium salt was isolated by addition of aqueous calcium acetate to the sodium salt **43** produced in this step. Hence, in the process development route, the final stages from the two components **18** and **19** were achieved without use of hazardous reagents, and these steps were maintained with minor modification in the final manufacturing route.

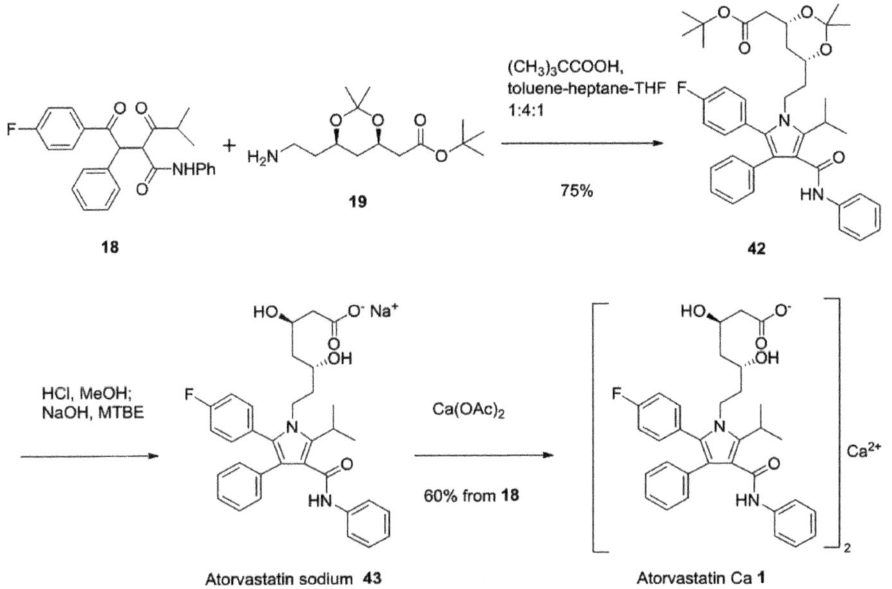

Scheme 4.12 Process development route – final stages.

4.4 The Final Manufacturing Route Involving Non-hazardous Reagents

4.4.1 Strategy

The overall strategy of the development route was highly successful, and was carried into manufacturing. As discussed in Section 4.3.3, synthesis of the side-chain component featured the highest number of hazardous reagents. Therefore, this had the greatest potential for improvement, and has been the focus of the majority of innovation.

4.4.2 1,4-Diketone Component 18

The route to this compound (Scheme 4.4) was changed relatively little. The loading of the relatively expensive catalyst for the Stetter reaction was reduced to 7 mol% and the reaction solvent was changed to isopropanol.[38] A problem with this route is release of benzaldehyde from the benzylidine adduct **21** in the presence of water and hence formation of the des-fluoro impurity, which is difficult to purge, in the Stetter reaction. A more convergent approach to the diketone **18** employing Friedel–Crafts acylation of fluorobenzene with phenylacetyl chloride, then halogenation of either ketone **44** or the isobutyryl acetanilide **20** followed by alkylation with the other component under basic conditions has been reported (Scheme 4.13).[39–41] While this may make control of the impurity easier, in requiring aluminium

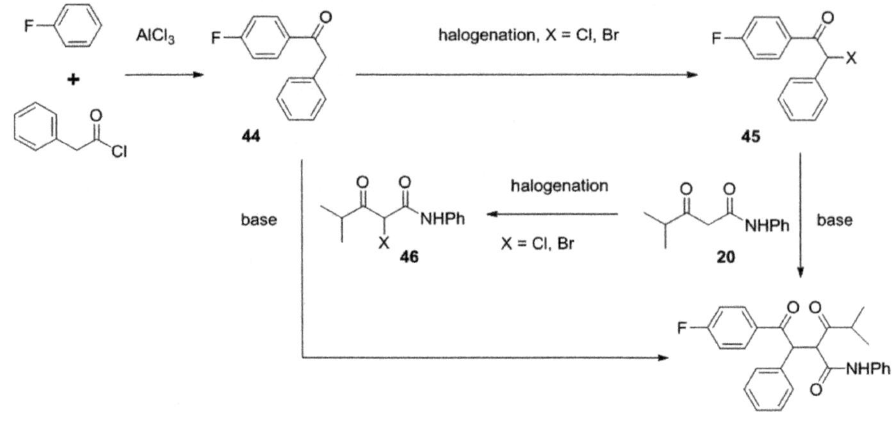

Scheme 4.13 Alkylation approach to diketone **18**.

chloride and halogenating reagents, and involving likely genotoxic alpha-halocarbonyl compounds, it utilizes a significantly greater number of hazardous reagents and intermediates.

4.4.3 Side-chain Component 19

Improvements to the development route have been driven by advances in catalysis, especially by implementation of biocatalysis for the introduction of the stereocentres (Scheme 4.14). Thus, ethyl (*S*)-4-chloro-3-hydroxybutyrate **48**, which can be produced in very high enantioselectivity by enzymatic reduction of readily available ethyl (*S*)-4-chloro-3-acetoacetate **47**, has become the preferred 5-carbon starting material,[20,42,43] avoiding the use of boron-based reductants, hydrogen peroxide and hydrobromic acid required in earlier chiral pool based routes. Isopropanol is cheaper than glucose, and can also be used as a reductant with certain enzymes;[44,45] in a further economic benefit with these enzymes, the glucose dehydrogenase for cofactor recycling is not required, and only a single enzyme is necessary. Asymmetric hydrogenation with ruthenium-based catalysts can also be used for this asymmetric reduction[46,47] and also proceeds in very high enantiomeric excess, but the biocatalytic process is more economically attractive. In the subsequent cyanation step, by using a halohydrin dehydrogenase enzyme,[42,48–50] allowing the reaction to be run at lower pH, and hence avoiding base-induced side reactions of the epoxide intermediate, higher yields are obtained. However, a consequence of the lower pH is the release of free hydrogen cyanide in the reaction medium and hence the headspace. Use of continuous cyanation[43,51] is another approach to improving the yield in this step, which avoids the hazard of free HCN, but does not give such high yields as the enzymatic process. In the subsequent Claisen condensation step, while LDA has been used on large scale, lithium bis(trimethylsilyl)amide

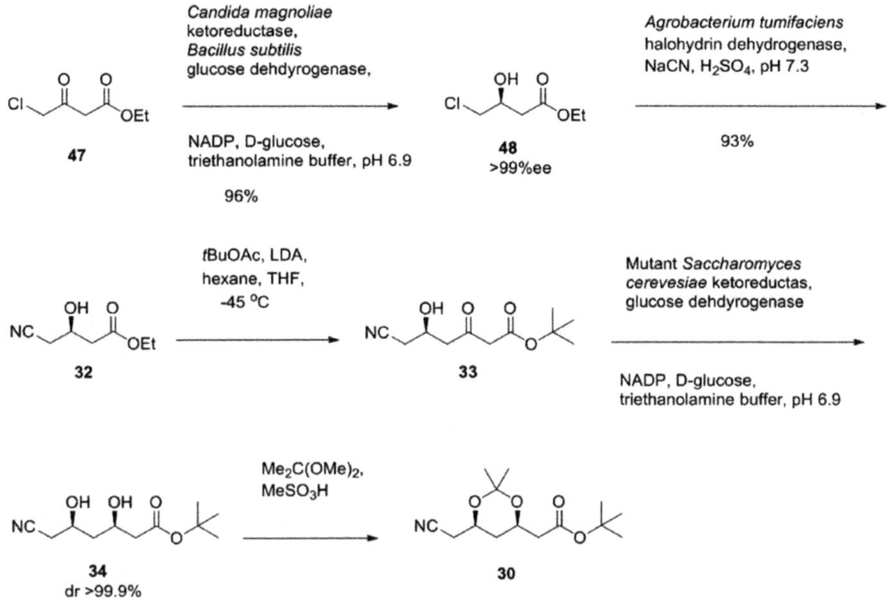

Scheme 4.14 Biocatalysis-based route to side-chain nitrile **30**.

(LHMDS), which does not require alkyllithium reagents for its generation, has also been used.[52] The replacement of the boron-based stereoselective reduction of keto-alcohol **30** with a more benign biocatalytic method has proved challenging, but by 2006, engineered enzymes capable of both high stereoselectivity and productivity in this step had been developed.[53–56]

The nitrile **30** is regarded as a key starting material for atorvastatin, and hence the final nitrile hydrogenation step in the development route (Scheme 4.7), which has been subject to no fundamental changes, strictly forms part of the endgame.

4.4.4 Endgame

In the endgame, the use of hazardous reagents was minimised at the development stage, and only minor modification was required for the transition to manufacturing. As discussed, owing to the hindered nature of diketone **18**, the Paal–Knorr condensation of diketone **18** and amine **19** with loss of a molecule of water is a slow reaction. This reaction is mechanistically complex and involves a number of reaction intermediates and equilibria.[57] The three-component solvent system was simplified to a 2-component system of THF and heptane.[58] Despite significant effort to find improved catalysts by both Pfizer and generics companies,[57,59] no major improvement over pivalic acid as a catalyst has been found, although salts of this acid such as the salt with trimethylamine offered some benefit. Carrying the reaction out at reduced pressure to enhance the rate by removal of water may also have some benefit.[60] The deprotection and isolation are carried out with slight modification to the isolation procedure,[57] in which no intermediates are isolated. However, yield and purity improvements may be possible by use of isopropyl alcohol instead of methanol for the acid deprotection step and isolation of the intermediate atorvastatin *tert*-butyl ester **17**.[61]

4.5 Conclusions

The nature of the reagents used in the synthesis of atorvastatin has been driven firstly by the change in requirements of each phase of development, and secondly by advances in technology for stereoselective ketone reduction. The improvements garnered over three generations of routes to atorvastatin are summarised in Table 4.1.

Thus, in medicinal chemistry, methodology allowing easy variation of substituents in the pyrrole core for study of structure-activity relationship and reliable introduction of the side-chain stereocentres was used. This involved a chiral auxiliary mediated acetate aldol reaction for introduction of the first stereocentre and alkylboron-based stereoselective ketone reduction methodology for the second, both requiring several hazardous reagents. In process development, the synthesis was re-ordered to achieve higher convergence, with final assembly of the pyrrole core moved to the endgame. Alternative methodology was also used for pyrrole construction, eliminating some hazardous reagents, but this re-ordering led to a requirement for use of cyanide, Raney nickel catalyst and ammonia for introduction of the nitrogen atom into the side-chain. For the side-chain, rather than asymmetric aldol chemistry, chiral pool sources for the first stereocentre were used, eliminating further hazardous reagents, but the alkylboron-based stereoselective ketone reduction remained for the introduction of the second stereocentre. In manufacturing, in the side-chain synthesis, employment of stereoselective biocatalytic ketone reduction allowed elimination of the hazardous reagents previously used in functional group interconversion of

Table 4.1 Summary of hazardous reagents used in routes to atorvastatin.

	Med-chem	Development	Manufacturing	Hazard type
Core	Bromo-ester			Toxic
	Isobutyryl chloride			Reactive, volatile
Side-Chain	N/A	H_2O_2		Oxidant
		HBr		Volatile acid
		LDA	LDA/LHMDS	Reactive
		NaCN	NaCN	Cyanide
		Et_2BOMe		Pyrophoric
Endgame	LDA			Reactive
	Et_3B			Pyrophoric
	H_2O_2			Oxidant
		Raney Nickel	Raney Nickel	Pyrophoric
		Ammonia	Ammonia	Toxic gas

chiral pool starting materials and substitution of those used in the stereoselective ketone reduction with non-hazardous D-glucose. Future development of existing chemistry as discussed in Section 4.3.3[36] or new chemistry may witness the demise of the remaining hazardous reagents required in the production of this important drug.

Acknowledgements

I am grateful to Drs Joanna Fox and Peter McCormack for checking and suggesting improvements to the text.

References

1. Z. Casar, *Curr. Org. Chem.*, 2010, **14**, 816.
2. A. Graul and J. Castaner, *J. Drugs Future*, 1997, **22**, 956.
3. B. D. Roth, *Prog. Med. Chem.*, 2002, **40**, 1.
4. D. J. Ager and C. R. Bayley, *Handbook of Chiral Chemicals*, ed. D. J. Ager, Taylor & Francis, Boca Raton, Florida, USA, 2nd edn, 2006, vol. 31, pp. 591–594.
5. B. D. Roth, C. J. Blankley, A. W. Chucholowski, E. Ferguson, M. L. Hoefle, D. F. Ortwine, C. S. Sekerke, D. R. Sliskovic, C. D. Stratton and M. W. Wilson, *J. Med. Chem.*, 1991, **34**, 357.
6. J. M. Lopchuk and G. W. Gribble, *Tetrahedron Lett.*, 2015, **56**, 3208.
7. (a) K. M. Chen, K. G. Gunderson, E. G. Hardtmann, K. Prasad, O. Repic and M. J. Shapiro, *Chem. Lett.*, 1987, 1923; (b) K. Narasaka and F.-C. Pai, *Tetrahedron*, 1984, **40**, 2233.
8. (a) M. Iqbal, N. Mistry and P. A. Clarke, *Tetrahedron*, 2011, **67**, 4960–4966; (b) Q. Xu, J. Yu, F. Han, Fengyan, J. Hu, W. Chen and L. Yang, *Tetrahedron: Asymmetry*, 2010, **21**, 156; A. Soriente, M. De Rosa, M. Stanzione, R. Villano and A. Scettri, *Tetrahedron: Asymmetry*, 2001, **12**, 959.
9. (a) A. Fettes and E. M. Carreira, *J. Org. Chem.*, 2003, **68**, 9274; (b) J. Krueger and E. M. Carreira, *J. Am. Chem. Soc.*, 1998, **120**, 837; (c) B. L. Pagenkopf, J. Kruger, A. Stojanovic and E. M. Carreira, *Angew. Chem., Int. Ed.*, 1998, **37**, 3124.
10. C. Jimenez-Gonzalez, C. Ollech, W. Pyrz, D. Hughes, Q. B. Broxterman and N. Bhathela, *Org. Process Res. Dev.*, 2013, **17**, 239.
11. C. Jimenez-Gonzalez, C. S. Ponder, Q. B. Broxterman, J. B. Manley and B. Julie, *Org. Process Res. Dev.*, 2011, **15**, 912.
12. K. L. Baumann, D. E. Butler, C. F. Deering, K. E. Mennen, A. Millar, T. N. Nanninga, C. W. Palmer, B. D. Roth and D. Bruce, *Tetrahedron Lett.*, 1992, **33**, 2283.
13. A. T. McKenzie, *Eur. Pat.* 849704, 1995; *Chem. Abstr.* 126, 199448.
14. C. A. Briggs, R. A. Jennings, R. A. Wade, K. Harasawa, S. Ichikawa, K. Minohara and S. Nakagawa, *Eur. Pat.* 848705, 1995; *Chem. Abstr.* 126, 212037.

15. M. Lin and D. Schweiss, *U. S. Pat.* 6274740, 1995; *Chem. Abstr.* 126, 199450.
16. D. E. Butler, T. V. Le, A. Millar and T. N. Nanninga, *U. S. Pat.* 5155251, 1991; *Chem. Abstr.* 118:41188.
17. P. L. Brower, D. E. Butler, C. F. Deering, T. V. Le, A. Millar, T. N. Nanninga and B. D. Roth, *Tetrahedron Lett.*, 1992, **33**, 2279.
18. Y. Wu, D.-J. Xiong and F.-E. Chen, *Tetrahedron*, 2015, **15**, 3487.
19. S. E. Bode, M. Wolberg and M. Mueller, *Synthesis*, 2006, 557.
20. M. Müller, *Angew. Chem., Int Ed.*, 2005, **44**, 362.
21. S. Hayashi, N. Ueyama, K. Inoue, T. Koga and S. Takahashi, *U. S. Pat.* 5,214,197, 1991; *Chem. Abstr.* 116:193719.
22. K. M. Chen, G. E. Hardtmann, G. T. Lee, J. Linder, S. Wattanasin and P. K. Kapa, *Eur. Pat.* 244364, 1987; *Chem. Abstr.* 108:131038.
23. K. Inoue, S. Takahashi, K. Yonetsu and N. Ueyama, *U. S. Pat.* 4,983,759, 1988; *Chem. Abstr.* 114:61543.
24. G. Beck, H. Jendralla and K. Kesseler, *Synthesis*, 1995, 1014.
25. N. Kizaki, Y. Yamada, Y. Yasohara, A. Nishiyama, M. Miyazaki, M. Mitsuda, T. Kondo, N. Ueyama and K. Inoue, *U. S. Pat.* 6903225, 1998; *Chem. Abstr.* 132:166230.
26. R. A. Holt, J. A. Blacker and C. D. Reeve, *U. S. Pat.* 7732171, 2001; *Chem. Abstr.* 135:356851.
27. S. Saito, T. Hasegawa, M. Inaba, R. Nishida, T. Fujii, S. Nomizu and T. Moriwake, *Chem. Lett.*, 1984, 1389.
28. K. Bock, I. Lundt and C. Pedersen, *Acta Chem. Scand. B*, 1983, **37**, 341.
29. M. Sletzinger, T. R. Verhoeven, R. P. Volante, J. M. McNamara, E. G. Corley and T. M. H. Liu, *Tetrahedron Lett.*, 1985, **26**, 2951.
30. R. P. Volante, T. R. Verhoeven, M. Sletzinger, J. M. Mcnamara, T. M. Liu and E. G. Corley, *U. S. Pat.* 4,582,914, 1985; *Chem. Abstr.* 105, 42566.
31. D. S. Karanewsky, *U. S. Pat.* 5025000, 1990; *Chem. Abstr.*, 115, 256384.
32. G. Wang and R. I. Hollingsworth, *Tetrahedron: Asymmetry*, 1999, **10**, 1895.
33. T. E. Jacks and D. E. Butler, *U. S. Pat.* 5998633, 1996; *Chem. Abstr.* 128:154074.
34. R. L. Hollingsworth, *U. S. Pat.* 5,292,939, 1991; *Chem. Abstr.* 118:168701.
35. G. Huang and R. I. Hollingsworth, *Tetrahedron*, 1998, **54**, 1355.
36. S. Hu, J. Tao and Z. Xie, *PCT Pat. Appl.* WO 2006134482, 2005; *Chem. Abstr.* 146:80524.
37. S. Werkmeister, K. Junger and M. Beller, *Org. Process Res. Dev.*, 2014, **18**, 289–302.
38. S. O'Sullivan and E. Fox, *PCT Pat. Appl.* WO 2007029217, 2005; *Chem. Abstr.* 146:316683.
39. B. Jiang, Y. Liu, Y. Huang, Y. Gao and P. Shen, *Chin. Pat. Appl.* CN 1299811, 2000; *Chem. Abstr.* 137:140336.
40. K.-U. Schoening and J. Hartwig, *PCT Pat. Appl.* WO 2003004457, 2001; *Chem. Abstr.* 138:106503.
41. J. Mathew, T. T. Puthiaparampil, M. Sridharan, S. P. Sathya, Shanker and S. Ganesh, *PCT Int. Appl.* WO 2004108660, 2003; *Chem. Abstr.* 142:37924.

42. S. K. Ma, J. Gruber, C. Davis, L. Newman, D. Gray, A. Wang, J. Grate, G. W. Huisman and R. A. Sheldon, *Green Chem.*, 2010, **12**, 81.
43. L. Proctor, P. J. Dunn, J. M. Hawkins, A. S. Wells and M. T. Williams in *Green Chemistry in the Pharmaceutical Industry,* ed. P. J. Dunn, A. S. Wells and M. T. Williams, Wiley-VCH, Weinheim, 2010, vol. 11, pp. 221–242.
44. H. Yamamoto, A. Matsuyama, Akinobu and Y. Kobayashi, *Biosci., Biotechnol. Biochem.*, 2002, **66**, 481.
45. N. Itoh, K. Isotani, M. Nakamura, K. Inoue, Y. Isogai and Y. Makino, *Appl. Microbiol. Biotechnol.*, 2012, **93**, 1075.
46. T. Saito, T. Yokozawa, T. Ishizaki, T. Moroi, N. Sayo, T. Miura and H. Kumobayashi, *Adv. Synth. Catal.*, 2001, **343**, 264–267.
47. W. Fan, W. Li, X. Ma, X. Tao, X. Li, Y. Yao, X. Xie and Z. Zhang, *J. Org. Chem.*, 2011, **76**, 9444.
48. R. J. Fox, C. S. Davis, E. C. Mundorff, L. M. Newman, V. Gavrilovic, S. K. Ma, S. L. M. Chung, C. Ching, S. Tam, S. Muley, J. Grate, J. Gruber, J. C. Whitman, R. A. Sheldon and G. W. Huisman, *Nat. Biotechnol.*, 2007, **25**, 338.
49. S.-Y. Chen, C.-X. Yang, J.-P. Wu, G. Xu and L.-R. Yang, *Adv. Synth. Catal.*, 2013, **355**, 3179.
50. P. Yao, L. Wang, J. Yuan, L. Cheng, R. Jia, M. Xie, J. Feng, M. Wang, Q. Wu and D. Zhu, *ChemCatChem*, 2015, **7**, 1438.
51. L. D. Proctor and A. J. Warr, *U. S. Pat.* 7531683, 2006; *Chem. Abstr.* 139:395631.
52. P. McCormack, A. J. Warr and E. Latham, *U. S. Pat. Appl.* US 20140051869, 2009; *Chem. Abstr.* 154:512923.
53. L. J. Giver, L. M. Newman and E. Mundorff, *U. S. Pat.* 8273547, 2006; *Chem. Abstr.* 148:420222.
54. M. P. Burns and M. Paul, Wong, John Wing, *PCT Pat. Appl.* WO 2008059366, 2006; *Chem. Abstr.* 148, 584146.
55. X. Wu and X. Gou, *Process Biochem.*, 2015, **50**, 104.
56. Further information related to this process is revealed in the following reviews (a) M. P. Burns, *Green Chem.*, 2013, **15**, 3099–3104; (b) A. S. Wells, J. W. Wong, P. C. Michels, D. A. Entwistle, K. Fandrick, G. L. Finch, A. Goswami, H. Lee, S. Mix, T. S. Moody, L. Pang, R. K. Sato, N. J. Turner and T. J. Watson, *Org. Process Res. Dev.*, 2016, **20**, 594–560.
57. S. Davies, A. Denholm, P. Kelleher and W. E. Tully, *Eur. Pat.* 1,922,315, 2005; *Chem. Abstr.* 145:356634.
58. S. O'Sullivan and J. O'Neill, *Eur. Pat.* 1,861,354, 2005; *Chem. Abstr.* 146, 316682.
59. M. K. Palaniswamy, S. R. Madarapu, V. H. Dahanukar, K. K. Machavaram, S. R. Poreddy, N. R. Tamma, S. C. Boyillapalli and R. B. Chhadua, *Indian Pat. Appl.* 2010CH03687, 2010; *Chem. Abstr.* 162, 259072.
60. B. De Lange, D. Heemskerk and H. L. M. Elsenberg, *PCT Pat. Appl.* WO 2014041058, 2013; *Chem. Abstr.* 160, 443871.
61. Y. V. Novozhilov, M. V. Dorogov, M. V. Blumina, A. V. Smirnov and M. Krasavin, *Chem. Central J.*, 2015, **9**, 7.

CHAPTER 5

Synthesis of Raloxifene

JAYA PANDEY

Amity School of Applied Sciences, Amity University Uttar Pradesh, Lucknow Campus 226028, Uttar Pradesh, India
Email: jpandey@lko.amity.edu

5.1 Synthesis of Raloxifene

Raloxifene, as shown in Figure 5.1, is a classic example of a selective estrogen receptor modulator (SERM)[1–7] and a number of studies indicate that the binding affinity of raloxifene is about one third of the binding affinity of estradiol[8,9] to the estrogen receptor. It is an estrogen agonist/antagonist, commonly referred to as a SERM that has benzothiopheneheterocyclic moiety.

It an extensively investigated SERM that partially simulates the effect of estrogens in bone and the cardiovascular system, while functioning as an anti-estrogen in endometrial and breast tissues. It has been approved for the prevention of osteoporosis and reduction in risk of fragility fracture. Thus, by administering an estrogen antagonist, an effect of functional deficiency of an estrogen is also created in the body, depriving the body all the beneficial effects of the hormone. Thus the prime requisite of targeted drug discovery is to achieve a critical balance of agonist/antagonist action in a molecule that may minimize or eliminate estrogenic activity in non-targeting tissues, optimize beneficial estrogenic effects in targeted tissues and eliminate the formation of carcinogenic metabolites. The current approach in the development of newer therapeutic agents is to identify compounds that may elicit tissue-specific effects. Many antiestrogens like EM-800, CP-336156, arzoxifene, raloxifene *etc.* are being developed as SERMs, showing agonist actions at the central nervous system, cardiovascular and skeletal systems, while

Green Chemistry Series No. 52
Hazardous Reagent Substitution: A Pharmaceutical Perspective
Edited by Rakesh Kumar Sharma and Rakeshwar Bandichhor

Published by the Royal Society of Chemistry, www.rsc.org

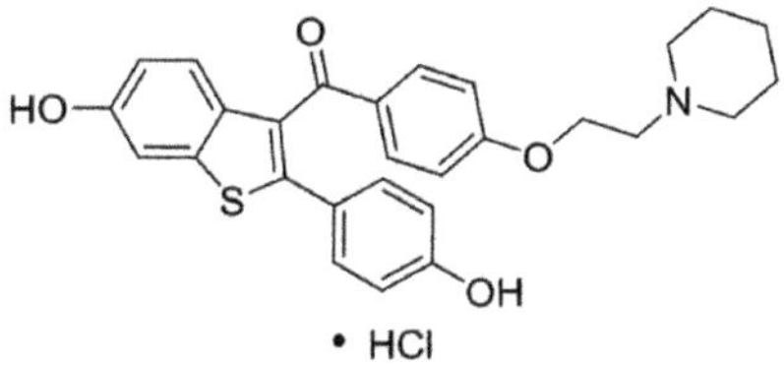

Figure 5.1 Raloxifene hydrochloride **1**.

antagonist action at the breast and uterus. The clinical success of these molecules for the treatment of estrogen-dependent diseases, such as osteoporosis, coronary artery disease, depression, Alzheimer's disease *etc.*, coupled with the novel finding of tissue specificity, has intrigued chemists and biologists to think about developing such agents. Recent exciting discoveries related to the structure and function of estrogen receptors has greatly enhanced the prospects for the development of novel ligands that may function as tissue-SERMs. As a new era has been entered into with the demonstration of multifunctional biological profiles of synthetic non-steroidal SERM, in the not too distant future there will be a selection of modulators available. The choice of modulators will be tailored to meet the needs of specific tissues and the risk profile of the individual therapy, providing better treatment options for endocrine-dependent disorders. Noticeably, this class of compounds shows promise for the treatment and prophylactic pathologies associated with estrogen, by which novel estrogen pharmaceuticals can be developed as tissue-selective drugs in the new millennium.

The search for ligands that can confer tissue-selective effects showing advantages of estrogens on non-traditional target tissues while mitigating some of the disadvantages, particularly concerns over estrogen agonist action, led to the emergence of a novel group of compounds showing pounced subtype ER alpha and ER beta selective differences in binding affinity and transcriptional efficacy and elicit agonist/antagonist responses depending on target tissue and hormonal milieu. This awareness has led to the development of molecules which could antagonize the effects of estrogens on uterine and breast tissues, while mimicking the effects of estrogens on bone and the cardiovascular system.

5.2 Importance of Raloxifene

Raloxifene slows the re-sorption of bone by reducing the biochemical markers of bone turnover in the pre-menopausal population. Raloxifene hydrochloride also lowers the chance of developing a certain type of breast cancer (invasive breast cancer) in post-menopausal women. These effects on bone are established as reductions of bone turnover markers in the serum and urine levels and a decrease in the incidence of fractures are observed.

Raloxifene, administered in a 60 mg dose once daily, increased spine and hip bone mineral density by 2 to 3%. This medicine was found to decrease

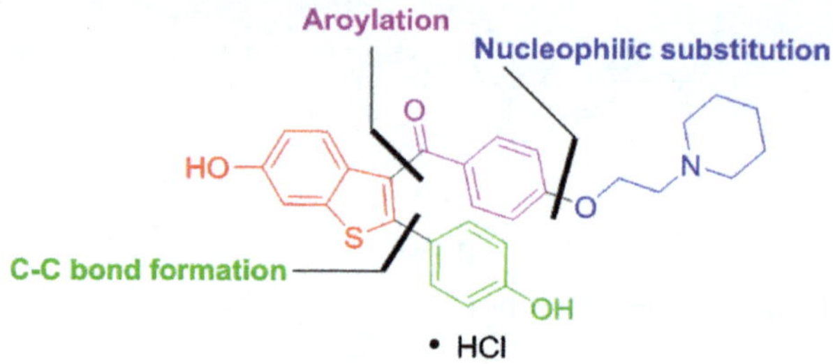

Scheme 5.1 Retrosynthetic analysis of **1**.

the incidence of the first vertebral fracture from 4.3% for placebo to 1.9%. As breast cancer is more commonly found in women over the age of 60, this group are maximally benefitted by raloxifene. It may also benefit women who have a family history of breast cancer or who have a genetic predisposition towards breast cancer. Other factors that can increase the risk of developing invasive breast cancer include a previous lobular carcinoma *in situ* diagnosis, a history of frequent breast biopsies, first child delivery at a later age, or starting the menstrual cycle at a later age.

Mechanistically, raloxifene works by blocking estrogen in the breast tissue of women at high risk. This medication helps prevent the spread of tumors that require estrogen to grow. It is found not to be effective in women who currently have invasive breast cancer or who have had invasive breast cancer. It neither prevents cancer in these women nor will it treat cancer once it appears.

Although it can reduce the likelihood of invasive breast cancer, it is important to be aware of its serious side effects *e.g.* increased chances of blood clots in the lungs and legs along with increased chance of stroke in women with coronary artery disease. The US Food and Drug Administration advises that women with the following conditions must not take raloxifene:

1. Those are pregnant or planning to become pregnant,
2. Individuals with present or past blood clots in the eyes, lungs, or legs,
3. People on cholestyramine, a cholesterol-lowering drug,
4. Those are on supplemental estrogen, and
5. Individuals are in pre-menopausal stage.

The structural architecture of raloxifene incorporates a benzothiophenyl unit carbon tethered basic side chain.

There are various methodologies, as shown in Scheme 5.1, featuring retrosynthetic analysis that exist for the synthesis of benzothiophenes and most of them employ nucleophilic substitution, Friedel–Crafts acylation (aroylation), and Grignard addition (C–C bond formation) as the key steps.[10,11]

5.3 First-generation Synthesis

Raloxifene is marketed as Evista by Eli Lilly which is an oral SERM prescribed to patients. The medicinal chemistry route[2] involves nucleophilic substitution in the presence of KOH, where 3-methoxybenzenethiolate **3** with 2-bromoacetophenone derivative **2** generate a thioether **4**. PPA mediated electrophilic cyclization and rearrangement then furnish the desired 6-methoxy-2-(4-methoxy-phenyl)-benzo[b]-thiophene **5** and an undesired 4-methoxy-2-(4-methoxy-phenyl)-benzo[b]-thiophene (not shown) as a mixture of regioisomers (para–ortho) in a 3 : 1 ratio. However, Vincenzi and co-workers optimized the reaction parameters and synthesized **5** with enhanced yields (98 : 2 para–ortho isomers).[12] The modified protocol utilized a solid acid catalyst in the presence of methanesulfonic acid. Treatment of **5** with *p*-alkoxy benzoyl chloride **7** (synthesized from corresponding acid **6**) in the presence of a Lewis acid *e.g.* $AlCl_3$ afforded the raloxifene free base **8** that can be converted to **1** as shown in Scheme 5.2.

Synthesis of radiolabelled raloxifene free base **11** is purposefully designed for probing the estrogen receptor and starts with intermediate **5** which reacts

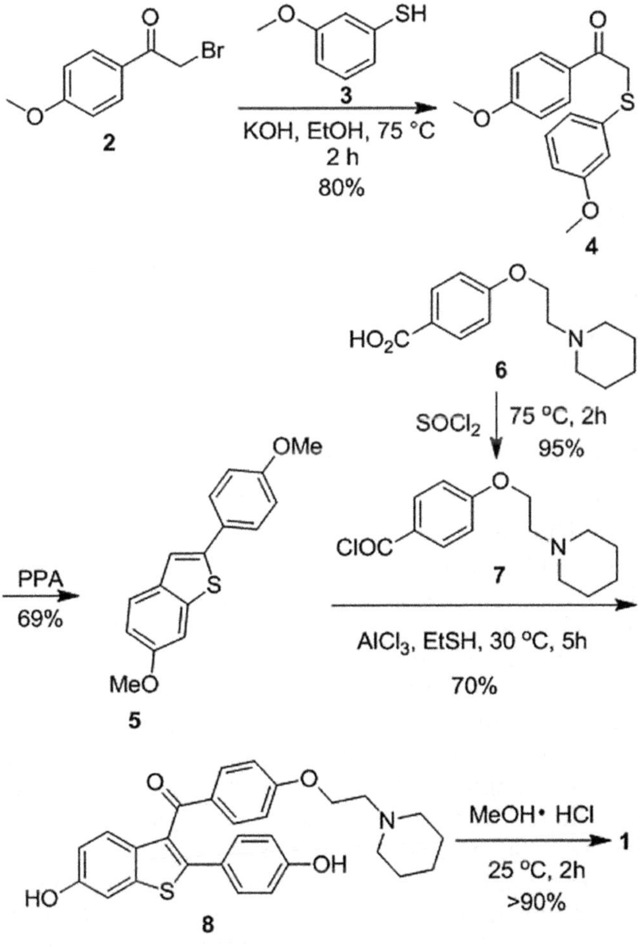

Scheme 5.2 First-generation synthesis.

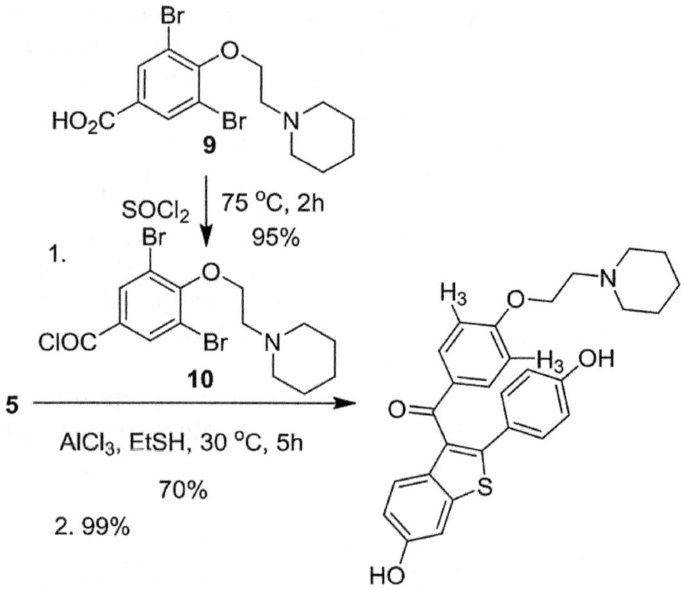

Radiolebelled 1 (free base)
11

Scheme 5.3 Synthesis of radiolabelled raloxifene free base **11**.

with aroyl chloride **10** (synthesized from corresponding acid **9**) in the presence of a Lewis acid *e.g.* $AlCl_3$. This affords a radiolabelled raloxifene intermediate (not shown) and as a result of didebromination this gets converted to free base **11** as shown in Scheme 5.3.[13]

5.4 Second-generation Synthesis

There were two new synthetic methodologies for raloxifene **1** that have been reported. The first one involves the acylation of *N*-(6-methoxy-1-benzothiophen-2-yl)-*N*,*N*-dimethylamine **15** with 4-fluorobenzoyl chloride **17** (synthesized from corresponding acid **16**) by heating at 100 °C in chlorobenzene, which affords the 3-acyl derivative **18**. Moreover, substrate **15** was synthesized by a reaction of the *p*-methoxy benzaldehyde **12** and *N*,*N*-dimethylthioformamide **13** in dry THF which was added to a solution of LDA at −78 °C. The resultant solution was allowed to slowly warm to room temperature over 3 h. The reaction was quenched by adding saturated ammonium chloride solution (40 mL), and the organic layer was removed, dried ($MgSO_4$), filtered, and solvent evaporated to yield the crude α-hydroxy thioacetamide **14**. The desired product was purified by recrystallization from ethanol with an 83% yield. Thioacetamide derivative **14** was converted to intermediate **15** in the presence of MsOH. Substrate **17** was synthesized by the addition of an excess of thionyl chloride to the carboxylic acid

16, heating under stirring for 2 h at 75 °C, removal of the remaining thionyl chloride by distillation at 760 Torr and fractionated distillation of the product (150 mm Vigreux column) under vacuum and usual workup yielded the product with 95% yield. Thereafter intermediate **15** is then reacted with aroylchloride **17** to afford intermediate **18.** Intermediate **18** is subjected with *p*-methoxyphenylmagnesium bromide in THF yielding 3-(4-fluorobenzoyl)-6-methoxy-2-(4-methoxyphenyl)-1-benzothiophene **19.** This is a very interesting C–C bond formation step where the Grignard reagent can potentially add to both carbonyls as well as at the 2nd position. Considering the presence of a Michael acceptor by virtue of the double bond present in the thiophene ring, the Grignard reagent preferably adds to the 2nd position yielding intermediate **19** with 83% yield. The condensation of **19** with 1-(2-hydroxyethyl)piperidine **20** by means of NaH in DMF affords the ether **21** with 90% yield, which was finally demethylated with $AlCl_3$ and ethanethiol to obtain a free base. Eventually, methanolic HCl was employed to obtain **1** as shown in Scheme 5.4.[14]

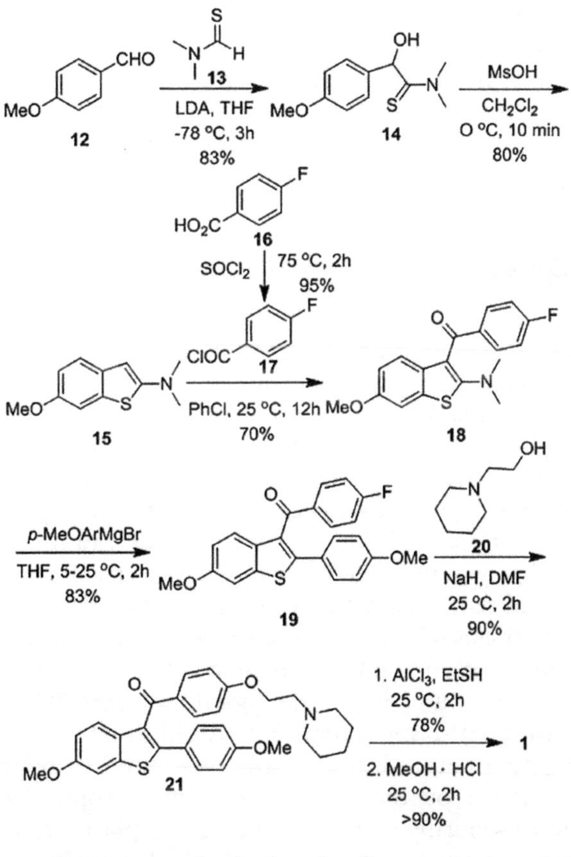

Scheme 5.4 Second-generation synthesis using first-generation synthesis including a Grignard reagent at an early stage.

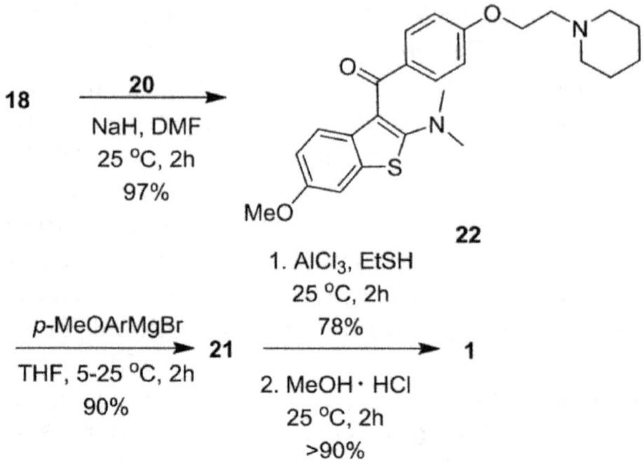

Scheme 5.5 Second-generation synthesis using a Grignard reagent at a late stage.

In the second strategy, the intermediate **18** can also be condensed first with 1-(2-hydroxyethyl)piperidine **20** by means of NaH as mentioned in Scheme 5.4 giving the piperidinoethyl ether **22**, which is then condensed with the *p*-methoxyphenylmagnesium bromide affording the ether **22** which was finally demethylated with $AlCl_3$ and ethanethiol to obtain a free base. Eventually, methanolic HCl was employed to obtain **1** as shown in Scheme 5.5.[14]

5.5 Third-generation Synthesis

In third-generation synthesis, ionic liquids (ILs) as greener alternatives have been used. ILs are considered to be organic salts that exist as liquids at low temperatures. The salient features of ILs are to have very low vapour pressures. Considering this fact, they are possibly categorized as green solvents in comparison to traditional volatile organic solvents. ILs have many attractive attributes such as chemical and thermal stability, high ionic conductivity *etc.* Therefore, they have been extensively investigated as solvents in various transformations including organocatalysis and inorganic synthesis, bio-catalysis and polymerization. ILs were found to have various applications in well-established reactions *e.g.* alkylations, allylations, hydroformylations, epoxidations, synthesis of ethers, Friedel–Craft reaction, Diel–Alder reaction, Knoevengal condensation and Wittig reaction. These solvents are considered perfect greener alternatives to volatile organic solvents. ILs are categorically better than traditional solvents (particularly chlorinated ones) not only due to their non-volatile nature but they are also found to be non-flammable. An example is given by the development of an optimized process for degreasing and/or scouring metal, ceramic, glass, plastic composite material or semiconductor surface, by treatment of the surfaces in a solution comprising an IL.

There are some ILs that have been the subject of toxicity and ecotoxicity studies and data are now available on a larger variety of organisms (both micro and macro-organisms). Most of the toxicological studies have been carried out on imidazolium- and pyridinium-based ILs, with alkyl or alkoxy side chains. Moreover, the anion counterparts studied are limited mainly to bromide, chloride, hexafluorophosphate and tetrafluoroborate. Nevertheless, a little research has been devoted to the determination of the biodegradability of the ILs and a prototype design of biodegradable ILs has recently been disclosed. A high throughput-screening-based diffusion method was recently employed to test the toxicity of ILs towards microorganisms and to distinguish toxic and incompatible ILs in one of the approaches.

Synthesis of raloxifene using ionic liquid solvents has been developed as shown in Scheme 5.6. Raloxifene is prepared starting from the 6-methoxy-benzothiophene-2 boronic acid **23**. This synthesis involves two carbon–carbon bond forming reactions, which are carried out by Suzuki coupling

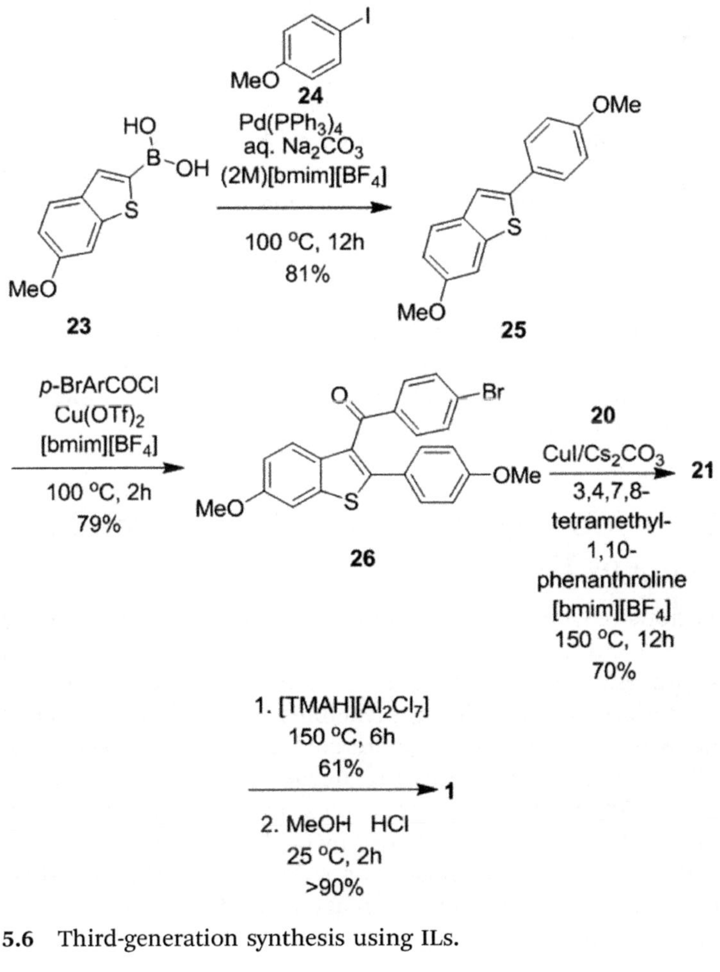

Scheme 5.6 Third-generation synthesis using ILs.

and Friedel–Crafts acylation reactions respectively. In this synthetic endeavor, Suzuki coupling reactions have been studied extensively in ILs using various halo and other substituted aromatic boronic acids as the starting material. As a part of the synthesis, 6-methoxybenzothiophene-2-boronic acid **23** is treated with 4-iodoanisole **24** in the presence of Na_2CO_3 and [bmim][BF4] (butylmethylimidazolium tetrafluoroborate), at 110 °C to obtain 6-methoxy-2-(4-methoxyphenyl)benzo[b]thiophene **25** with a 81% yield. The product was extracted with solvent toluene. The toluene is removed from the product by distillation under reduced pressure. The ionic liquid is recovered by washing it with water to remove Na_2CO_3. After successful Suzuki coupling, a second step was performed by using Friedel–Crafts acylation reaction with 4 bromobenzoylchloride in the IL [bmim][BF4] and a catalytic amount of $Cu(OTf)_2$ to furnish **26**. The product is extracted with toluene and the IL is recovered after successive washing with water. The next step is the introduction of 2-hydroxyethyl piperidine side chain **20** on substrate **26**.

Particularly, this reaction has been carried out in a dipolar aprotic solvent such as DMF or DMSO using strong bases such as NaH, or alkoxides. Both these conditions suffer from the disadvantage that the removal of the solvent to separate the product is difficult. In addition, NaH and alkoxide bases are highly hygroscopic and cannot be used with an imidazolium-based IL.[15] Subsequently, substrate **26** is treated with 2-hydroxyethyl piperidine **20** in the presence of copper(I) iodide, 3,4,7,8-tetramethyl-1,10- phenanthroline, and Cs_2CO_3 in [bmim][PF6] at 150 °C to yield 70% of **21**. Copper and Cs_2CO_3 free ionic liquid [bmim][PF6] is recovered after product isolation using toluene. The penultimate step to achieve the target is carried out in the presence of the Lewis acidic ionic liquid [TMAH][Al_2Cl_7] (trimethylammonium aluminium chloride) with co-solvent methylene chloride to yield the raloxifene free base (not shown) at 61% yield.[16] Eventually, methanolic HCl was employed to obtain **1** as shown in Scheme 5.6.

5.6 Transitioning from Hazardous to Non-hazardous Reagent Use

Non-hazardous reagents are to be substituted for hazardous reagents where possible, to avoid generating hazardous waste. During the execution of chemical processes, significant quantities of waste are unavoidable. There are many initiatives and guiding principles in the literature but daunting hazardous waste-related challenges remain unaddressed. Hazardous waste by definition creates an everlasting negative impact on health and the environment. Green chemistry principles are inclusive of safer and non-hazardous substrates and reagent use by design.[17]

It is quite evident that in most drug substance synthesis (using the medicinal chemistry route) the starting reagents and substrates are hazardous. Thereafter, the synthesis of drug substances evolves and employs relatively

Table 5.1 Transitioning from hazardous to non-hazardous reagent use.[a]

	Use of reagents		
Schemes	Aroylation	C–C bond formation	Nucleophilic substitution
5.1	—	—	—
5.2	**$AlCl_3$**	**PPA**	**NaH**
5.3	**$AlCl_3$**	**PPA**	**NaH**
5.4	*PhCl*	*ArMgBr*	*NaH*
5.5	*PhCl*	*ArMgBr*	*NaH*
5.6	$Cu(OTf)_2$, IL	$Pd(PPh_3)_4$, ArMgBr, IL	CuI, Cs_2CO_3, IL

[a]Bold: hazardous reagents; italic: semi-hazardous reagents; upright: non-hazardous reagents' combination; IL: ionic liquid.

much safer practices and less hazardous reagents. Deliberate efforts towards finding non-hazardous reagent substitutes during synthesis design may offer potential solutions to environmental issues caused by the pharmaceutical industry to a great extent.

For instance, the raloxifene synthesis landscape reveals the route evolving over time from the medicinal chemistry strategy to third-generation synthesis. First- and second-generation synthesis involves the usage of hazardous reagents ($AlCl_3$, NaH), whereas in the third-generation synthesis all the strategic steps, *e.g.* aroylation, C–C bond formation and nucleophilic substitution, employ metal-catalyzed or mediated transformation as shown in Table 5.1. The solution is not ideal but it is safer than first- and second-generation syntheses.

5.7 Conclusion

It would be impossible to list all of the possible chemical hazards which might be encountered in academic laboratories and industry. There are a few commonly used materials that can be considered hazardous but finding suitable substitutes is nearly impossible. For instance, nitric acid, which is corrosive in nature, and its oxides are found to be highly toxic. In general, nitric acid is a powerful oxidizing agent and forms flammable and explosive compounds with many materials. In addition to this, perchloric acid formulates highly explosive and unstable materials with many flammable materials and with a few metals. When combined, perchloric and sulfuric acids form anhydrous perchloric acid that can ignite any organic matter instantaneously. It is also reported that crystals may form in perchloric acid storage containers which might explode if stored for more than one year. Picric acid may also form explosive compounds with many combustible materials. Wet storage is considered to be safe, particularly if it is stored away from combustible materials, and it should not be kept for extended periods. Hydrofluoric acid is corrosive and this is the only acid that even attacks glass. This is very toxic and it can damage nerves too. It readily reacts with metals or ammonia and forms toxic fumes. The vapour pressure of mercury at room temperature is very low and the heating of it must be avoided as it can pose a high chemical hazard to humans at higher temperatures; several times more than the allowable limits. Mercury and its derivatives should never be disposed of in the sewers. Lead is very commonly encountered during experiments and it gives toxic vapours when heated. Yellow phosphorus is recommended to be stored in water due to the fact that it may ignite instantaneously if allowed to dry. It is also found to be extremely toxic if ingested. Open-chain or cyclic ethers form peroxides if stored under atmospheric conditions. These peroxides are prone to explode from shock or even from the minor friction *e.g.* when opening the bottle. Carbon disulfide is a very peculiar solvent and its vapour can be ignited by contact with an electric light bulb or hot steam radiator. Active metals like Na and K are stored under oil due to the fact that they react violently with water and may ignite spontaneously if exposed to the moisture in air. Long-term standing K, even in oil, forms a superoxide. The dry superoxide may explode if subjected to any form of shock.

Oxidizing agent and peroxide packing should be in minimum quantities as they may deteriorate after long storage causing an explosion. Organic phosphates are found to be so toxic that these compounds were used in making in nerve gases used during World War II. They are also used as pesticides. These are a few examples and there are many such hazardous reagents used in laboratory practice. It may be possible to design safer reagents for chemical transformations up to some extent but there are some trivial yet extremely important and unavoidable reagents with which we may have to live with and use them with utmost safety and precautions as described above.

In summary, in this chapter we provided a select overview of practical synthetic strategies towards the synthesis of raloxifene HCl **1**. Recent advances in the synthesis of **1** are outlined in detail, with classification into three generations: (1) first-generation synthesis; (2) second-generation synthesis; and (3) third-generation synthesis.

In general, raloxifene HCl **1** utilizes three broad strategies (aroylation, C–C bond formation and nucleophilic substitution) where 6-methoxy-2-(4-methoxy-phenyl)-benzothiophene is used as core building block. All three generation syntheses have been probed considering the use of hazardous reagents. It was found that the third-generation synthesis employs comparatively less hazardous reagents.

Acknowledgements

Support from the Department of Chemistry, Amity University Lucknow Campus is greatly acknowledged. The author is also thankful for financial support to DST (Grant registration 236/2013) and UPCST (Grant registration # CST/ D 6547/2017).

References

1. T. A. Grese and J. A. Dodge, *Curr. Pharm. Des.*, 1998, **4**, 71–92.
2. H. U. Bryant and W. H. Dere, *Proc. Soc. Exp. Biol. Med.*, 1998, **217**, 45–52.
3. C. D. Jones, M. G. Jevikar, A. J. Pike, M. K. Peters, L. J. Black, A. R. Thompson, J. F. Falcone and J. A. Clemens, *J. Med. Chem.*, 1984, **27**, 1057–1066.
4. M. Sato, T. A. Grese, J. A. Dodge, H. U. Bryant and C. H. Turner, *J. Med. Chem.*, 1999, **42**, 1–24.
5. M. W. Draper, D. E. Flowers, W. J. Huster, J. A. Neild, K. D. Harper and C. J. Arnaud, *J. Bone Miner. Res.*, 1996, **11**, 835–842.
6. S. R. Cummings, *J. Am. Med. Assoc.*, 1999, **281**, 2189–2197.
7. T. Hol, M. B. Cox, H. U. Bryant and M. W. J. Draper, *J. Womens Health*, 1997, **6**, 523–531.
8. C. R. Schmid and J. P. Sluka, *Tetrahedron Lett.*, 1999, **40**, 675.
9. T. A. Grese, S. Cho, D. A. Finkey, A. G. Godfrey, C. D. Jones, C. W. Lugar III, C. W. , M. J. Martin, K. Matsumoto, L. D. Pennington, M. A. Winter, M. D. Adrian, H. W. Cole, D. E. Magee, D. L. Phillips, E. R. Rowley, L. L. Short, A. L. Glasebrook and H. U. Bryant, *J. Med. Chem.*, 1997, **40**, 146.
10. R. A. Altman, A. Shafir, A. Choi, P. A. Lichtor and S. L. Buchwald, *J. Org. Chem.*, 2008, **73**, 285.
11. W. J. Smith III and J. S. Sawyer, *Tetrahedron Lett.*, 1996, **37**, 299.
12. J. T. Vincenzi, T. Y. Zhang, R. L. Robey and C. A. Alt, 3, *Org. Proc. Res. Dev.*, 1999, 56.
13. J. A. Dodge, M. G. Stocksdale and C. D. Jones, *J. Radiolabelled Compd. Radiopharm.*, 1995, **36**, 43.
14. D. A. Bradley, A. G. Godfrey and C. R. Schmid, *Tetrahedron. Lett.*, 1999, **40**, 5155.
15. P. L. Arnold, M. Rodden, K. M. Davis, A. C. Scarisbrick, A. J. Blake and C. Wilson, *Chem. Commun.*, 2004, **14**, 1612.
16. P. S. Shinde, S. S. Shinde, A. S. Renge, G. H. Patil, A. B. Rode and R. R. Pawar, *Lett. Org. Chem.*, 2009, **6**, 8.
17. *Green Chemistry: Theory and Practice*, ed. P. Anastas and J. C. Warner, Oxford University Press, Oxford, 1998.

CHAPTER 6

Synthesis of Montelukast

M. SARAVANAN, E. R. R. CHANDRASEKHAR,
VILAS H. DAHANUKAR AND RAKESHWAR BANDICHHOR*

IPDO-Innovation Plaza, Dr Reddy's Laboratories Limited, Bachupally,
Qutubullapur, RR District, Hyderabad, 500090, Telangana, India
*Email: rakeshwarb@drreddys.com

6.1 Introduction

Montelukast Sodium **1**, as shown in Figure 6.1, developed and marketed by Merck as Singulair® and launched in 1998, is probably one of the most commercially successful small molecule drugs ever launched. It is used for the management of asthma and seasonal allergies.[1–3] Structurally, montelukast sodium **1** has four distinct components. Three of the components contain heteroaromatic or aromatic rings (quinoline, meta-substituted phenyl, *ortho*-substituted phenyl) and are tethered by two-carbon and three-carbon chains to form a backbone. The fourth component, 2-(1-mercaptomethyl)cyclopropyl) acetic acid is linked to the backbone three-carbon chain to create the chiral center. A variety of synthetic approaches to montelukast sodium and its synthetic intermediates have been reported.[4–12]

The discovery and development of montelukast and approaches to its manufacture have been reviewed.[3] This overview traces the evolution of synthetic philosophy, which is inextricably linked with the available technology and reagents used in the transformations during the course of chemical development.

Green Chemistry Series No. 52
Hazardous Reagent Substitution: A Pharmaceutical Perspective
Edited by Rakesh Kumar Sharma and Rakeshwar Bandichhor

Published by the Royal Society of Chemistry, www.rsc.org

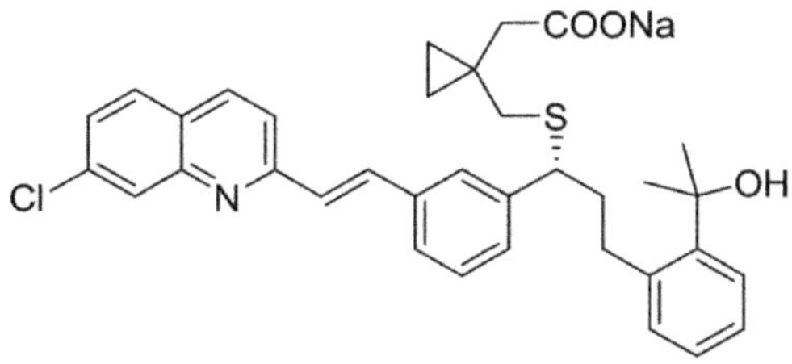

Figure 6.1 Structure of montelukast sodium **1**.

6.2 Medicinal Chemistry Route

6.2.1 Synthetic Strategy

In the discovery synthesis of montelukast sodium at Merck,[4] the most important feature of the synthetic route was the ability to vary substituents at the chiral center (thiol side chain and *ortho*-substituted phenyl) to establish the structure–activity relationship. The key features of the route are flexibility, to allow the introduction of a variety of substituents, and reliability, while ensuring a minimum number of steps from a common intermediate incorporating a thiol side chain and various substitutions on the phenyl ring. Though several routes were explored to access the variety of the compounds, most of the synthetic routes made use of diol **13** and thiol intermediate **20** for the construction of **1**.

6.2.2 Synthesis of Diol Intermediate 13

The synthesis of diol intermediate **13** (Scheme 6.1) begins by condensing 7-chloroquinaldine **2** with *m*-phthlaldehyde **3**. The quinoline methyl group is reacted with isophthalaldehyde in presence of acetic anhydride[13,14] and the resulting crude product is crystalized from ethyl acetate to afford aldehyde **4**. The addition of methyl magnesium bromide to aldehyde **4** provided the secondary alcohol **5**, which is then oxidized with activated manganese dioxide to afford ketone **6**. The enolate of ketone **6** is generated by reaction with KH or NaH, and reacted with dimethyl carbonate to provide the corresponding keto ester **7**.

Ortho-substituted phenyl group is incorporated by treatment of **7** with NaH followed by reaction with benzyl bromide **8** to furnish diester **9**, which is then decarboxylated with an acetic acid–hydrochloric acid mixture and then the resulting acid **10** is esterified with methyl iodide and potassium carbonate to give keto ester **11**. There are several strategies reported for the asymmetric reduction of **11**.[15–18] Reduction of **11** with chiral oxazaborolidine in the presence of borane/THF complex affords the chiral hydroxy ester **12**, which upon reaction with excess methyl lithium gave diol intermediate **13**.

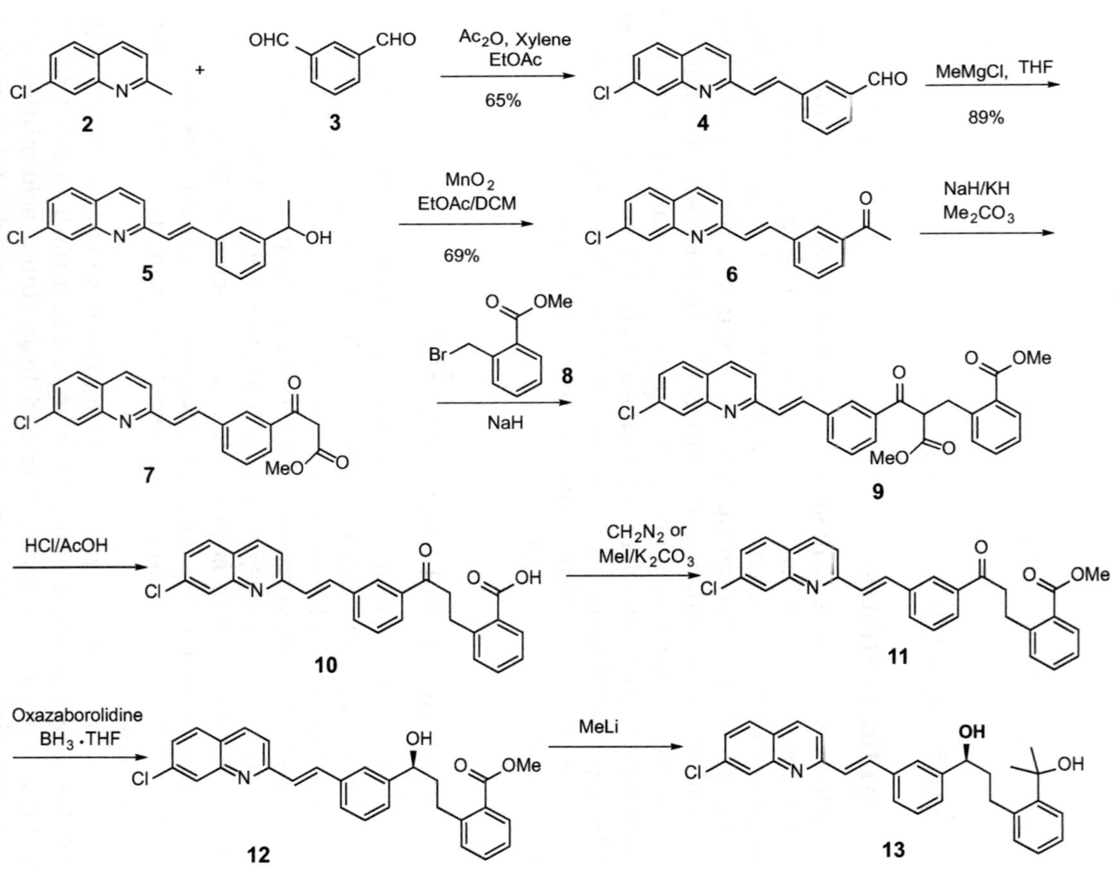

Scheme 6.1 Medicinal chemistry route to diol intermediate **13**.

6.2.3 Synthesis of Thiol Acid (20)

The thiol side chain **20** (Scheme 6.2) is accessed by reduction of commercially available diethyl 1,1-cyclopropane dicarboxylate **14** with $LiAH_4$ to provide alcohol **15**.[4,19–21] The objective is to selectively convert one of the two identical hydroxyl groups to nitrile and the other to thiol. To incorporate the nitrile function, initially alcohol is mono-protected as benzoate ester with benzoyl chloride and the other free alcohol **16** is mesylated with methanesulfonyl chloride followed by displacement with NaCN to incorporate the cyano functional group. Subsequently, cyanide **17** is hydrolysed and the resulting acid is methylated with diazomethane to get the hydroxy methyl ester **18**. Finally, the thiol group is introduced through mesylation of **18** with methanesulfonyl chloride and subsequent displacement with cesium thioacete. The thioacetate **19** thus obtained is treated with hydrazine and KOH yielded thiol acid side chain **20**.[4]

6.2.4 Synthesis of Montelukast (free acid of 1)

Once the two key building blocks **13** and **20** are created, the next step is the introduction of the thiol side chain in a stereoselective manner (Scheme 6.3).[4] To accomplish this, first secondary alcohol is protected as a silyl ether by reaction with *tert*-butylchlorodiphenylsilane and then the tertiary alcohol group is protected as a 2-tetrahydropyranyl (THP) ether derivative by reacting with 3,4-dihydro-2*H*-pyran and triphenylphosphine hydrobromide. Subsequent chromatographic purification yields the doubly-protected ether, which upon treatment with tetrabutylammonium fluoride leads to the formation of THP protected alcohol **21**. Mesylation of secondary alcohol followed by displacement with thiolate anion generated from reaction of **20** with NaH or caesium carbonate gives THP protected montelukast **23**. Finally, the hydrolysis of the THP group using pyridinium *p*-tolunesulfonate formed montelukast as a pale yellow solid (free acid form of **1**).

6.3 Process Development Route

6.3.1 Synthetic Strategy

The main drawbacks of the medicinal chemistry route are: (a) use of pyrophoric reagents ($LiAlH_4$, NaH, KH, MeLi, *etc.*); (b) higher number of steps; (c) protection and deprotection steps; (e) purification by column chromatography as well as repeated purifications; (f) lower yields with inferior quality which makes the process less attractive for large-scale synthesis. Hence, development of an alternative route or improving the existing route is indispensable to make the process more effective and scalable. As part of process research and development, a large number of innovative routes to montelukast and its intermediates have been reported.[5–12] Among all the

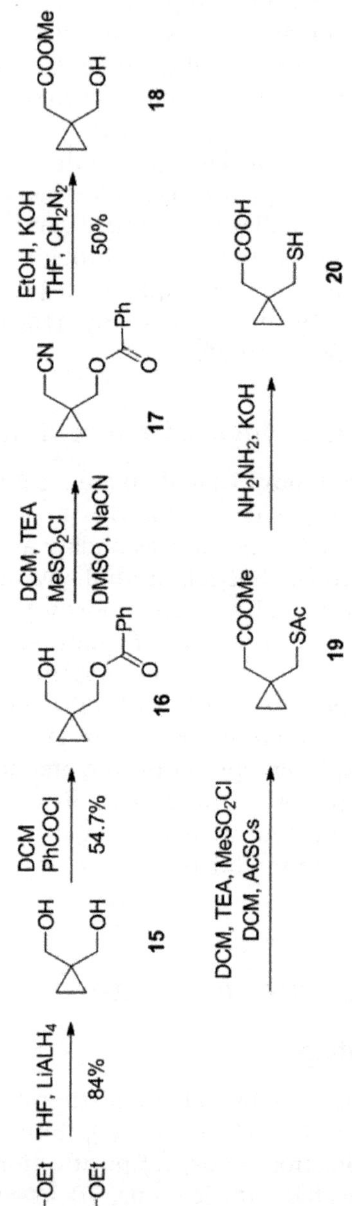

Scheme 6.2 Medicinal chemistry route to thiol acid **20**.

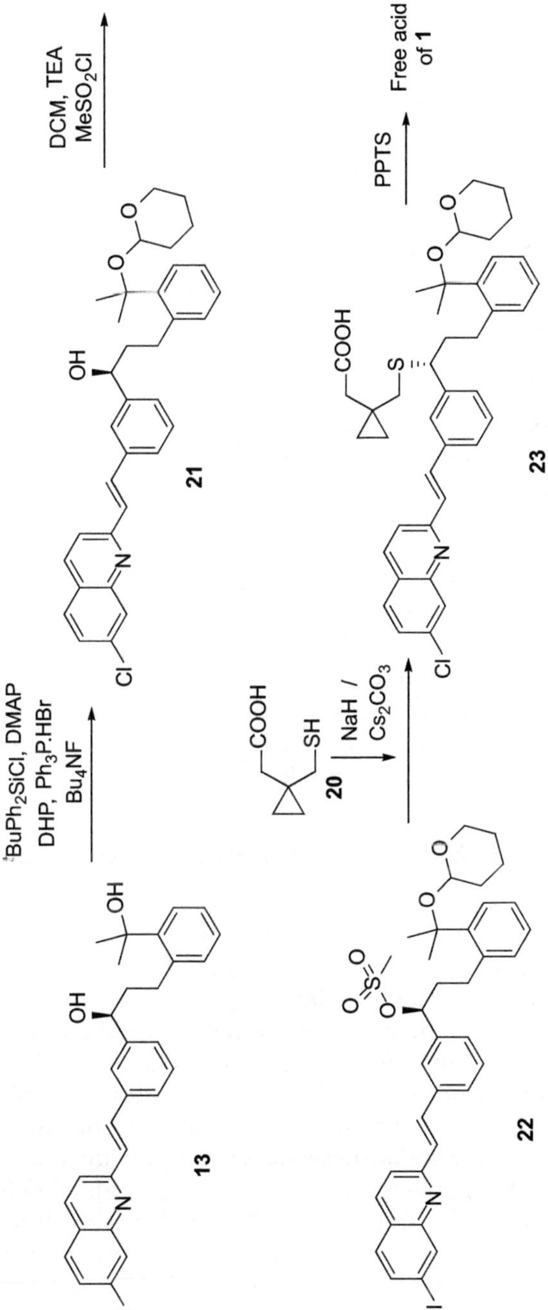

Scheme 6.3 Medicinal chemistry route to montelukast (free acid of **1**) from **13** and **20**.

synthetic strategies reported, studies directed towards development of more efficient and less hazardous routes for the efficient synthesis of two key building blocks, diol (**13**) and thiol acid (**20**), and their conversion into montelukast have been the area of focus for generic companies.[5] In search of these, significant advancements have been made in the synthesis of thiol acid and assembly of thiol acid and diol to montelukast. The diol route used in the process development followed the same general approach used in medicinal chemistry route as discussed in Section 6.2.2.

6.3.2 Synthesis of Diol 13 and Thiol Acid 20

Scheme 6.4 depicts the improved synthesis of thiol acid **20** starting from the commercially available 1,1-cyclopropanedimethaol **15**.[5] Unlike the medicinal chemistry route (Scheme 6.4), the nitrile group in is introduced by reacting **15** with thionyl chloride in the presence of diisopropylethylamine followed by the reaction of the resulting cyclic sulphite **24** with sodium cyanide and a catalytic amount of sodium iodide to provide the corresponding hydroxy-nitrile **25**. Thus, selective mono-protection of two hydroxyl groups using benzoyl chloride is avoided. With the hydroxynitrile compound now present, the introduction of the thiol group is accomplished *via* mesylation and subsequent reaction with potassium thioacetate in the presence of triethylamine. Finally, one step hydrolysis of thioacetate and nitrile functional groups with sodium hydroxide in biphasic medium yields the sodium salt of thiol acid **20** after removal of neutral impurities *via* extraction in toluene. Thiol **20** can be isolated after neutralization of the sodium salt as a pale brown solid or as a solution in toluene, thus, use of a biphasic reaction system avoids tedious chromatographic purification.

6.3.3 Synthesis of Montelukast Sodium 1 from 13 and 20

Finally, in the synthesis of **1**, protection and deprotection of the *tert*-alcohol group is avoided and the acid base salt formation approach is implemented in the montelukast free acid formation step (Scheme 6.5) to purge out impurities.[5] To achieve the preferential mesylation of secondary alcohol of the diol intermediate **13**, the molar equivalent of methanesulfonyl chloride is added at -25 ± 2 °C in a mixture of toluene-acetonitrile and diisopropylethylamine. The resulting reaction mass solution is seeded with the mesylate compound and the precipitated product is isolated by filtration to afford **27** as a pale yellow solid (mesylate **27** is highly unstable and needs to be stored below -20 °C). In another flask, dilithium salt of **20** is prepared by reacting **20** with *n*-butyl lithium at -15 ± 2 °C with tetrahydrofuran (THF) as the solvent. To the resulting dilithium salt solution, the solution of mesylate **27** in THF is added and aged at -5 ± 2 °C for 8.5 h. The reaction is quenched with ethyl acetate and 10% aqueous sodium chloride solution. The layers are separated and the organic layer is washed with 0.5 M tartaric acid solution and then concentrated to an oily residue. The residual montelukast free acid is dissolved in ethyl acetate and treated with dicyclohexylamine and the resulting solution is seeded with pure montelukast dicyclohexylamine salt. After the formation of a thick slurry, hexane is slowly added and the slurry aged at 20 ± 2 °C overnight. The precipitated product is filtered and dried to afford montelukast dicyclohexylamine salt (DCHA salt of **1**). Finally, the dicylohexylamine salt is neutralized with aqueous acetic acid in a mixture of toluene and water. The organic layer containing the free acid is treated with a solution of sodium hydroxide in 1% aqueous ethanol to ensure the clear dissolution of sodium salt and then concentrated to a minimum volume. The residual solution is dissolved in acetonitrile at 40 ± 2 °C and seeded with montelukast sodium crystalline salt. The resulting mixture is maintained at 40 ± 2 °C for 1.5 h, and then additional acetonitrile is added slowly and the slurry aged at 40 ± 2 °C for 12 h followed by room temperature for 1 h. Finally, the separated product is filtered and dried to afford montelukast sodium **1**.

6.4 Final Manufacturing Route Involving Non-hazardous Reagents

6.4.1 Synthetic Strategy

The overall strategy of the process development route was highly successful, and was carried into manufacturing. As discussed in Section 6.3.3 and 6.3.3, synthesis of the diol component follows linear synthetic methodology and featured the highest number of hazardous reagents; synthesis of downstream process (**13** to **1**) involves lengthy unit operations and makes the process less attractive at large scales. Therefore, these have the greatest potential for improvement, and have been the focus of the majority of process innovation.

6.4.2 Synthesis of Diol Intermediate

Improvements in the development route for diol **13** have been driven by advances in metal mediated cross coupling and catalysis. The overall synthetic efficiency was achieved through convergence by employing C–C bond formation between alkene **28** and halide **29**; and deploying homogeneous metal-based catalysis in the reduction of ketone to chiral diol **13** (Scheme 6.6). The route is started with the preparation aldehyde **4** as described in Section 6.2.2 followed by reacting the obtained aldehyde **4** with vinyl magnesium bromide in a mixture of toluene–THF (2 : 1) at 0–5 °C. After completion, the reaction is quenched with ammonium acetate and the layers were separated. The organic layer is concentrated to a residue and then the addition of hexanes provided allyl alcohol **28**.[22–24] Elaboration of allyl alcohol beautifully illustrates the power of the Heck arylation of alkenes in the synthesis. A tandem Heck reaction and double-bond isomerization between a allyl alcohol **28** and a hindered 2-(2-halophenyl)propan-2-ol **29** secured direct

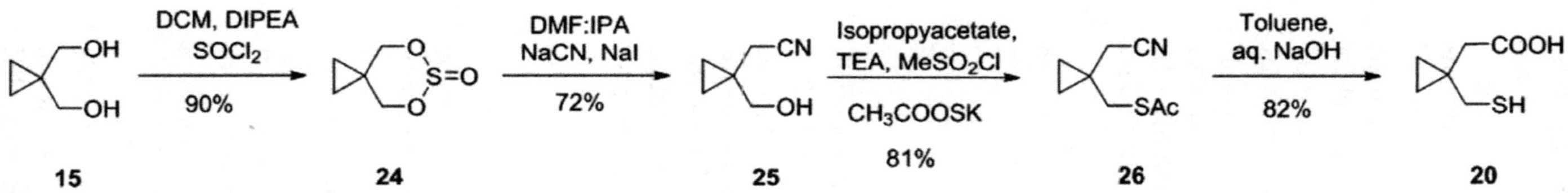

Scheme 6.4 Process development route to thiol acid (**20**).

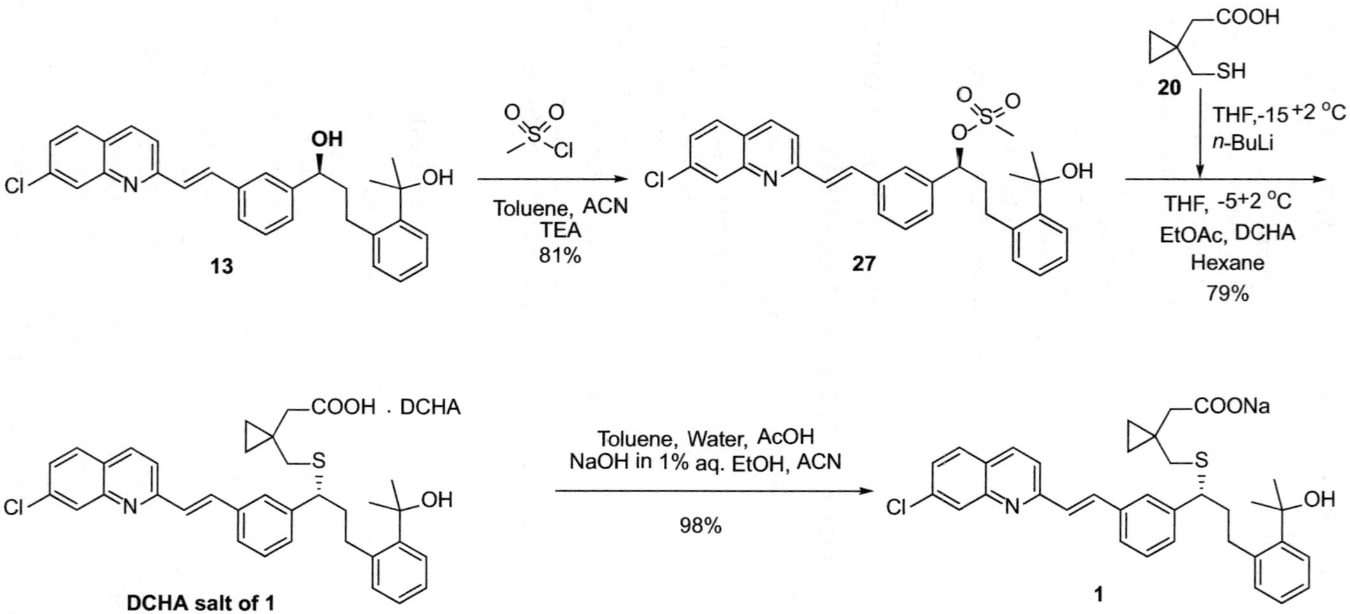

Scheme 6.5 Process development route to montelukast sodium **1**.

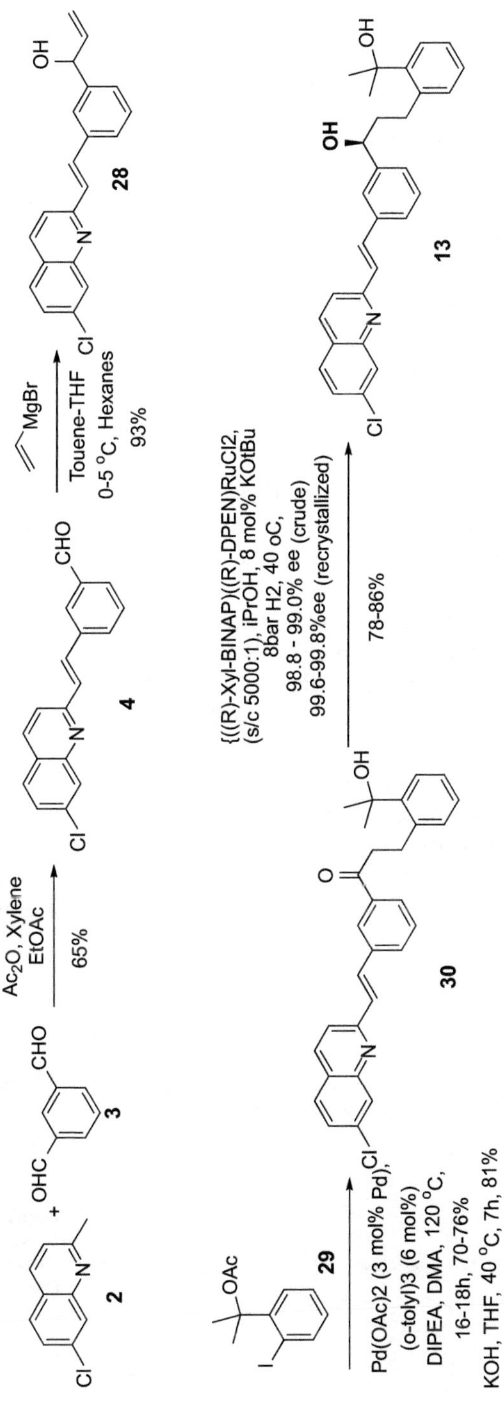

Scheme 6.6 Final manufacturing route to diol intermediate **13**.

access to the keto *tert*-alcohol **30**.[25] To a mixture of degassed DMA, **28** and **29**, is added $Pd(OAc)_2$, (*o*-tolyl)$_3$P and DIPEA. The reaction mixture is stirred at room temperature to dissolve the solids and obtain a clear orange-coloured solution which is gradually heated at 120 °C for 18 h. The product obtained after conventional work-up is dissolved in a mixture of THF and methanol, KOH is added and heated at 45 °C for 6 h to furnish keto *tert*-alcohol **30**. Finally, asymmetric hydrogenation of the ketone functionality in **30** provided a convenient method to introduce the benzylic alcohol chiral center. The hydrogenation vessel is charged with ketone **30** and isopropanol. The vessel is purged with nitrogen, then a solution of [((*R*)-Xyl-BINAP((*R,R*)-DPEN)$RuCl_2$] in deoxygenated isopropanol (5000 : 1) and potassium *tert*-butoxide are added. The vessel is pressurized to 7.6 bar with hydrogen and heated to 40 °C for 3.5 h. The reaction mixture is cooled to room temperature and quenched with acetic acid.

The resulting solution is concentrated to the desired volume and extracted into toluene. The toluene layer is seeded with **13** and then heptane is added to precipitate the product which is filtered and dried to yield diol **13** as a white solid with 86% yield and 99.6% ee.[25] Thus, a simple and convergent route is developed for **13** which avoids the usage of several hazardous reagents in the synthesis.

In another approach (Scheme 6.7), methyl 2-iodobenzoate **31** is used as a coupling partner in the Heck arylation with alkene. Coupling of allyl alcohol **28** with methyl 2-iodobenzoate, triethylamine, $Pd(OAc)_2$ in acetonitrile or toluene in reflux conditions affords a single step preparation of keto ester **11**.[10,22] The keto functionality in **11** is converted into the desired (*S*)-alcohol by asymmetric reduction with (−)-diisopinocamphenylchloroborane ((−)-DIPCl) in THF. The resultant hydroxyl ester **12** is subjected to a Grignard reaction with methyl magnesium chloride in toluene–THF in the presence cerium chloride. After quenching with dilute acetic acid and subsequent routine work-up, diol **13** is isolated by crystallization from toluene-hexanes.[18,24,26]

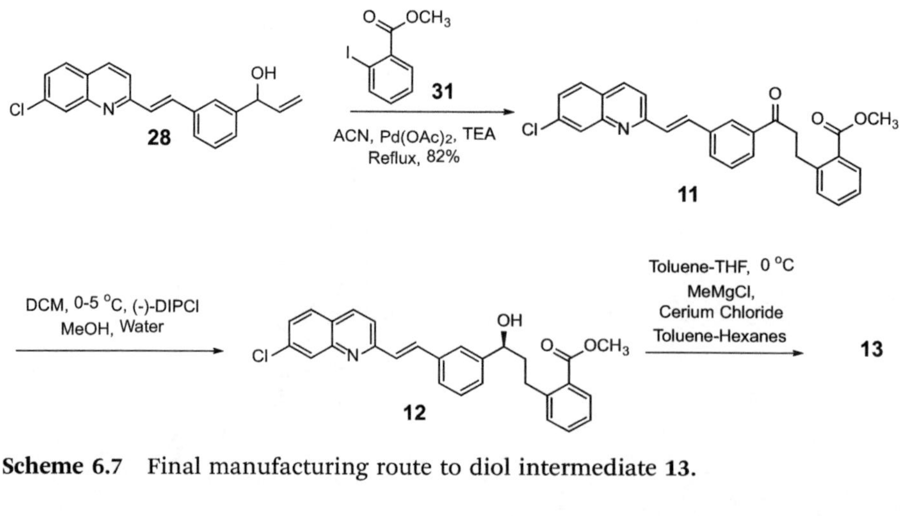

Scheme 6.7 Final manufacturing route to diol intermediate **13**.

6.4.3 Synthesis of Montelukast Sodium 1

With diol **13** and thiol acid **20** in hand, the next step is to activate the secondary alcohol as a leaving group and then displace the leaving group with thiolate with clean inversion of stereochemistry (Scheme 6.8). Mesylation could be achieved either as per the procedure described in Section 6.3.3 or alternatively reaction of **13** with methanesulfonyl chloride in DCM in the presence of diispropylethylamine at 12.5 ± 2.5 °C for 30–60 min, followed by the addition of hexane to enable the isolation of the product. The separated product is filtered and washed with hexane to afford the mesylated product (**27**, wet) as a pale yellow solid.[25b] During the filtration most of the process-related impurities (olefin, cyclic ether and **13**) are washed off in the DCM-hexane solvent system. The product, being unstable in nature, needs to be processed through the next step immediately to achieve a superior conversion. Experimental conditions suggest that either exposure to atmospheric air or temperature (>5 °C) and prolonged storage of the product leading to the formation of a predominant 7-membered cyclic ether impurity due to intramolecular displacement mesyl group by the hydroxyl group in *tert*-alcohol moiety. Dilithium salt of **20** is prepared separately by the reaction of **20** with *n*-butyl lithium at -15 ± 2 °C in THF, and wet **27** is added and aged at -5 ± 2 °C for 6 h. The reaction is quenched with aqueous ammonium chloride and layers are separated. The separated layer is concentrated to a residue and then dissolved in toluene.

The resulting solution is seeded with TBA salt of montelukast and aged for 20 h. Hexane is added to the stirred mass to get crystalline product. The separated product is filtered and the resulting wet solid is recrystallized with toluene to yield the TBA salt of montelukast with >99.5% purity. Finally, the TBA salt is neutralized with aqueous acetic acid in the presence of DCM and water. The organic layer containing the free acid is treated with a solution of sodium hydroxide in methanol and the resulting solution is subjected to evaporation using the Agitated Thin Film Distillation (ATFD) technique to

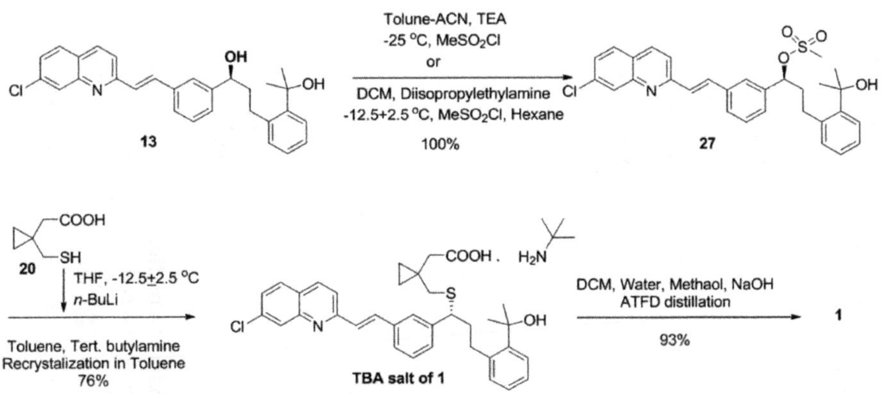

Scheme 6.8 Final manufacturing route to diol intermediate **13**.

Table 6.1 Comparative analysis of the synthesis routes.

Back bone	Medicinal chemistry	Development	Manufacturing	Hazard type
Diol intermediate	Acetic Anhydride	Acetic Anhydride	Acetic Anhydride	Reactive
	MeMgCl	MeMgCl	—	Reactive
	MnO_2	MnO_2	—	Oxidant
	NaH	NaH	—	Pyrophoric
	Me_2CO_3	Me_2CO_3	—	Flammable
	HCl/AcOH	HCl/AcOH	—	Volatile acid2
	CH_2N_2	CH_2N_2	—	Toxic, explosive
	MeI	MeI	—	Toxic, carcinogen
	Oxazaborolidine	Oxazaborolidine	—	—
	BH_3–THF	BH_3–THF	—	Flammable, vapour
	MeLi	MeLi	—	Pyrophoric
			Vinyl Magnesium bromide	Reactive
			$Pd(OAc)_2$	Precious metals considered environmentally burdensome unless recovery is established
			$(O\text{-tolyl})_3P$	
			$[((R)$-Xyl-BINAP$((R,R)$-DPEN$)RuCl_2]$	
Thiol acid side chain	$LiAlH_4$	—	—	Pyrophoric
	Benzoyl chloride	$SOCl_2$	$SOCl_2$	Reactive
	$MeSO_2Cl$	$MeSO_2Cl$	$MeSO_2Cl$	Reactive
	NaCN	NaCN	NaCN	Cyanide
	CH_2N_2	—	—	Toxic, explosive
	NH_2NH_2	—	—	Toxic
Elaboration of diol to Montelukast	$tBuPh_2SiCl$	—	—	Corrosive
	DHP	—	—	Flammable, toxic
	$Ph_3P \cdot HBr$	—	—	Corrosive
	Bu_4NF	—	—	Corrosive
	$MeSO_2Cl$	$MeSO_2Cl$	$MeSO_2Cl$	Reactive
	Cs_2CO_3	—	—	*n*-BuLi
	PPTS	—	—	Acute toxic
		n-BuLi	*n*-BuLi	Pyrophoric

yield motelukast sodium **1** as a white to off-white powder.[25b,27] Thus the reaction cycle time and work-up procedure are simplified to make the process suitable for large-scale manufacturing.

6.5 Summary

Montelukast sodium was accessed for the shortest synthetic route and the use of minimal hazardous reagents. Also, several unit operations are simplified at the manufacturing stage when compared to medicinal chemistry route. The improvements garnered over three generations of routes to montelukast sodium are summarized in Table 6.1. Future development of existing chemistry as discussed in Section 6.3.3[36] or new chemistry may witness the demise of the remaining hazardous reagents required in the production of this important drug.

As shown in Table 6.1, comparative analysis of the routes discussed reveal that the maximum hazardous reagents were used in medicinal chemistry route. Early development route features relatively less number of hazardous reagents whereas manufacturing route employs least number of reagents and few of them may be categorized as hazardous ones. Reagents like NaH, MeLi, and CH_2N_2 which have been used in the synthesis of diol intermediate are not safe to handle at large scales. There are few reagents, *e.g.* $LiAlH_4$ and CH_2N_2, which have been avoided in the thiol acid side chain involved in the early development route. Eventually, the route evolved from medicinal chemistry to manufacturing replacing most of the hazardous reagents.

Acknowledgements

We thank the management of Dr Reddy's Laboratories Ltd. (DRL) for supporting this work.

References

1. M. A. Lipkowitz and T. Navarra, *The Encyclopedia of Allergies Facts on File*, New York, 2nd edn, 2001, p. 178.
2. Merck prescribing information. www.merck.com/product/usa/pi_circulars/s/singulair/singulair_pi.pdf.
3. P. J. Harrington, *Pharmaceutical Process Chemistry for Synthesis: Rethinking the Routes to Scale-UP*, John Wiley & Sons, Hoboken, NJ, 2011, pp. 164–216.
4. (a) M. L. Belly, S. Leger, M. Labelle, P. Roy, Y. B. Xiang and D. Guay, *U. S. Pat.* 5565473, 1996; (b) M. L. Belly, S. Leger, M. Labelle, P. Roy, Y. B. Xiang and D. Guay, *E. P. Pat.* 408717 B1, 1998.
5. (a) M. Bhupathy, J. M. McNamara, D. R. Sidler, R. P. Volante and J. Bergan, *U. S. Pat.* 5614632, 1997; (b) M. Bhupathy, J. M. McNamara, D. R. Sidler, R. P. Volante and J. Bergan, *U. S. Pat.*, WO 9518107 A1, 1995.

6. (a) B. B. Jordi, R. J. Llorenc, P. A. Mireia, V. E. Xavier and G. C. Esther, *Eur. Pat.*, WO 2007/101841 A2, 2007; (b) S. King, B. Pipik, D. A. Colon, *U. S. Pat.* 5523477, 1996.
7. K. Danvor, K. Vensa and O. Renata, *Eur. Pat.*, WO 2008/049922 A2, 2008.
8. M. Saravanan, B. Satyanarayana and P. Pratap Reddy, *Synth. Commun.*, 2013, **43**, 2050–2056.
9. S. Venkataraman, T. R. Srinivasan, B. V. V. N. Chandra Sekhar, S. Alokkumar, K. Ravi Kumar and A. Sanjeev Kumar, *U. S. Pat.* 7189853 B2, 2007.
10. M. L. Belly, S. Leger, M. Labelle, P. Roy, Y. B. Xiang and D. Guay, *U. S. Pat.* 5565473, 1996, *U. S. Pat.* 5270324, 1993 and *Eur. Pat.* 480717 B1, 1998.
11. T. B. Bellmunt, J. L. Rafecas, A. M. Pasto and E. X. Verdaguer and C. E. Gordo, *Eur. Pat.*, WO 2007101841 A3, 2007.
12. D. Kidemet, V. Kroselj and R. Osolnik, *Eur. Pat.*, WO 2008049922 A3, 2008.
13. H. H. Tung and D. M. Hobbs, *U. S. Pat.* 5,869,673, 1999.
14. J. M. McNamara, J. L. Leazer, M. Bhupathy, J. S. Amato, R. A. Reamer, P. J. Reider and E. J. J. Grabowski, *J. Org. Chem.*, 1989, **54**, 3718.
15. M. M. Chartrain, T. C. Shieh-Shung, G. M. Garrity, B. Heim-buch, C. Roberge and A. Shafiee, *U. S. Pat.* 5,491,077, 1996.
16. S.-S. T. Chen and A. Shafiee, *U. S. Pat.* 5,427,933, 1995.
17. (a) A. Avdagic, J. Jerkovic and D. Pavlicic, *Br. Pat.*, WO 2008009970 A2, 2008; (b) A. Avdagic, B. Mohar, D. Sterk and M. Stephan, *Croatian Pat.*, WO 2006008562 A1, 2006; (c) A. Fujii, S. Hashiguchi, N. Uematsu, T. Ikariya and R. Noyori, *J. Am. Chem. Soc.*, 1996, **118**, 2521.
18. S. L. Pathi, R. N. Kankan, D. R. Rao and M. S. Phull, *Br. Pat.*, WO 2008032099 A2, 2008.
19. T. Imamoto, N. Takiyama, K. Nakamura, T. Hatajima and Y. Kamiya, *J. Am. Chem. Soc.*, 1999, **111**, 4392.
20. M. Labelle, M. Belley, Y. Gareau, J. Y. Gauthier, D. Guay, R. Gordon, S. G. Grossman, T. R. Jones, Y. Leblanc, M. McAuliffe, C. McFarlane, P. Masson, K. M. Metters, N. Ouimet, D. H. Patrick, H. Piechuta, C. Rochette, N. Sawyer, Y. B. Xiang, C. B. Pickett, A. W. Ford-Hutchinson, R. J. Zamboni and R. N. Young, *Bioorg. Med. Chem. Lett.*, 1995, **5**, 283.
21. J. M. L. Belley, *U. S. Pat.* 5,212,180, 1993.
22. A. O. King, E. G. Corley, R. K. Anderson, R. D. Larsen, T. R. Verhoeven, P. J. Reider, Y. B. Xiang, M. Belley, Y. Leblanc, M. Labelle, P. Prasit and R. J. Zamboni, *J. Org. Chem.*, 1993, **58**, 3731.
23. S. Suri, G. S. Sarin and M. Mahendru, *In. Pat.*, WO 2006021974 A1, 2006.
24. P. R. Pratap, B. Satyanarayana, N. Naidu, D. Ghosh, K. V. K. Kumar, J. R. Rajender, K. R. Kumar and M. Saravanan, *US Pat.*, WO 2008058118 A2, 2008.
25. (a) B. Satyanarayana, M. Saravanan, J. R. Rajender, K. V. K. Kumar, M. Laxman, S. Sen, D. A. Chaplin, R. C. Lloyd, M. Thomas, V. H. Dahanukar, O. Srinivas and M. E. Fox, *J. Org. Chem.*, 2015, **80**, 3891–3901; (b) M. E. Fox, B. Satyanarayana, M. Saravanan, J. R. Rajender and K. V. K. Kumar, *U. S. Pat.*, WO 2010/148209 A2, 2010.

26. (a) M. Bhupathy, D. R. Sidler, J. M. McNamara, R. P. Volante and J. Bergan, *U. S. Pat.* 6,320,052, 2001; (b) M. Bhupathy, D. R. Sidler, J. M. McNamara, R. P. Volante, J. Bergan, *U. S. Pat.* 5,614,632, 1997; (c) M. Bhupathy, J. M. McNamara, D. R. Sidler, R. P. Volante and J. Bergan, *U. S. Pat.*, WO 9518107 A1, 1995.
27. B. Satyanarayana, K. V. Kirankumar, M. V. Ramakrishna, K. R. Janardana Sarma and M. S. Kushal, *U. S. Pat.*, WO 2009/117381 A2, 2009.

CHAPTER 7

Development of a Safe, Scalable, Azide-free Synthesis of 1-Aryl-1H-tetrazoles Using Diformylhydrazine

SIVARAJ RAMASAMY,* THIRUMALAI LAKSHMINARASIMHAN, JAYAPRAKASH KARAMIL, SABUJ MUKHERJEE AND MANJUNATH GUJJAR

Chemical and Synthetic Development, Biocon Bristol-Myers Squibb Research and Development Center, Biocon Park, Jigani Link Road, Bommasandra IV, Bangalore 560099, India
*Email: Siva.Raj@syngeneintl.com

7.1 Introduction

Tetrazoles are nitrogen-rich compounds that have received considerable attention and have found a wide range of applications in synthetic and medicinal chemistry.[1] The first tetrazole-containing compound was reported by Bladin in 1885.[2] Since then, several methodologies have been developed to synthesize unique molecules of this class. Tetrazoles have not been found in nature. Except in rare cases, tetrazoles themselves do not exhibit appreciable biological activity. However, the tetrazole ring is incorporated as a key structural feature in several well-known drugs.[1,3] It has long been recognized that the tetrazole moiety (especially 5-substituted 1*H* tetrazoles) can serve as a surrogate for terminal carboxylic acids[4] and can often improve the

Green Chemistry Series No. 52
Hazardous Reagent Substitution: A Pharmaceutical Perspective
Edited by Rakesh Kumar Sharma and Rakeshwar Bandichhor

Published by the Royal Society of Chemistry, www.rsc.org

biological activity of the parent molecules.[5,6] Although tetrazoles and carboxylic acids are structurally different from each other, they display similar physico-chemical properties. Hansch reported that the lipophilicity of the anionic tetrazole is 10 times higher than that of the corresponding carboxylate, and this could account for its higher membrane permeability.[6]

The advantage of tetrazoles over carboxylic acids is their resistance to metabolic degradation[7] and their superior hydrogen bonding capability. The ability of all four nitrogen atoms in tetrazoles to act as hydrogen bonding acceptors can either enhance or reduce receptor-substrate interactions[8] – this property has been exploited by several drug discovery teams.

The presence of basic nitrogen atoms in tetrazoles allows them to exist as zwitterions – a phenomenon that could reduce the adsorption of potential drug candidates. To overcome this, the acidic tetrazole proton can be masked with a 'bioreversible' functional group. This strategy has been employed to increase the bioavailability of several molecules.[9]

This motif is not only used in pharmaceuticals but has also found utility in materials science as specialty explosives,[10] rocket fuel and as a gas-generating agent for air bags.[11] It is evident that this heteroaromatic ring system with four nitrogen atoms can form stable complexes with metal ions. Consequently, tetrazoles have been used in coordination chemistry as efficient complexes in corrosion inhibitors[12] and as intermediates for a variety of synthetic transformations.[13]

Because of the increasing utility of tetrazoles in multiple areas, it is important to develop safe and scalable methods for their preparation. The high energy associated with these molecules and their synthetic precursors makes this task a daunting one.

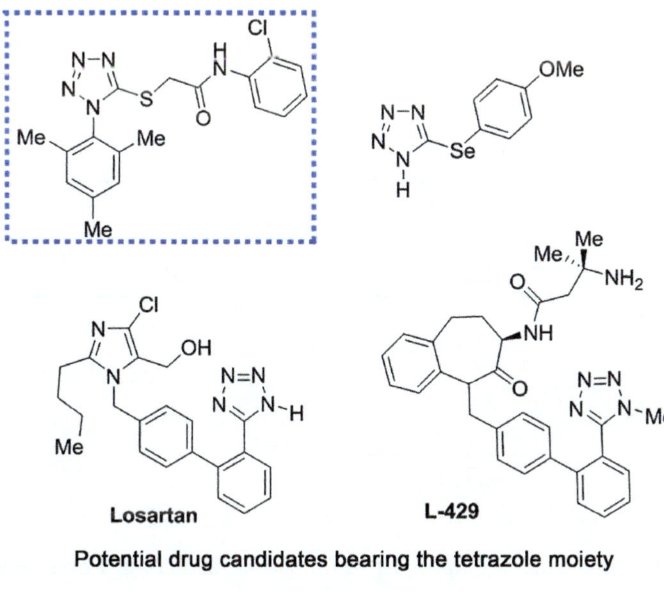

Potential drug candidates bearing the tetrazole moiety

In a recent program we were required to prepare kilogram quantities of 1-(4-chloro-3-fluoro-2-iodophenyl)-1*H*-tetrazole, **1**. Several strategies have been developed for the synthesis of 1-alkyl or 1-aryltetrazoles,[14] and some representative approaches are depicted in Scheme 7.1. Approach 'a' relies on the reaction of aryl isocyanides with hydrazoic acid.[15] Alternatively, the target tetrazoles may be accessed from isothiocyanates *via* reaction with hydrazoic acid, followed by subsequent oxidative desulfurization of the intermediate thiol (approach 'b').[16] The von Braun method (approach 'c') involves treatment of substituted amides or *N*-arylbenzamides with phosphorous pentachloride, followed by the reaction of the resultant imide chloride with hydrazoic acid to afford the desired tetrazole.[17] In all of these approaches, the coupling partners (isocyanide in approach 'a', isothiocyanate in approach 'b', and the imide chloride in approach 'c') are generally not readily available, and have to be accessed from the corresponding amines. This limitation adversely affects the atom economy and the efficiency of these approaches.

Approach 'd', which was our initial approach as well, relies on a three-component coupling reaction between an aryl amine, trialkyl orthoformate, and sodium azide in the presence of either a protic acid (acetic acid or trifluoroacetic acid),[18] or a Lewis acid (such as $Yb(OTf)_3$),[19] to furnish the desired 1-aryltetrazole. We adopted this method to provide gram quantities of **1** *via* reaction of 4-chloro-3-fluoro-2-iodoaniline and **2** with sodium azide and triethyl orthoformate in glacial acetic acid. This transformation was effected at 90 °C, and delivered the product in 50% yield after column chromatography (Scheme 7.2).

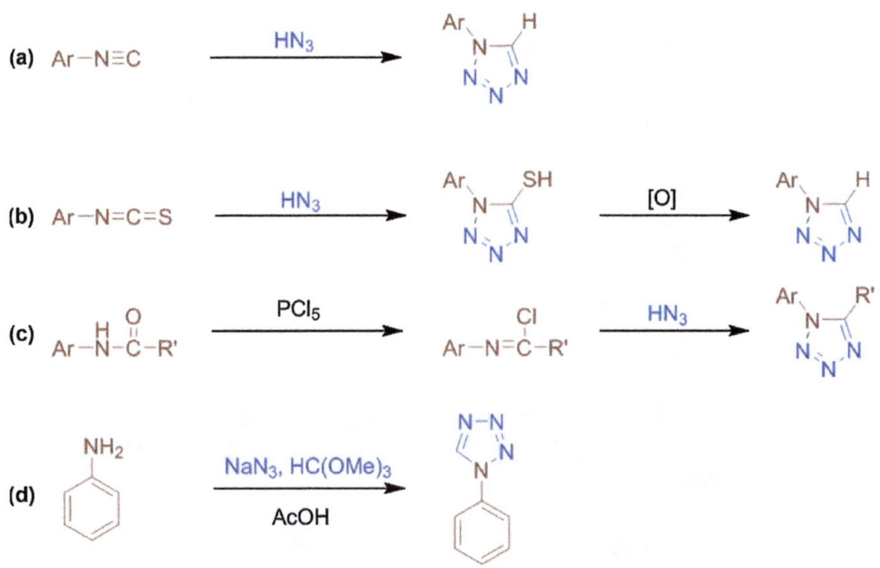

Scheme 7.1 Representative approaches for synthesis of 1-alkyl or 1-aryltetrazoles.

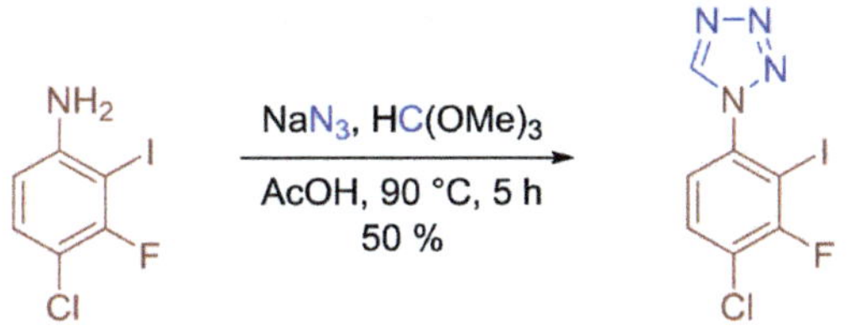

Scheme 7.2 Initial Synthesis of **1**.

While this approach is atom economical and can provide the tetrazole in one step from readily available starting materials, it is still fraught with the same limitations as the other three methods ('a', 'b' and 'c'), namely, the use of either hydrazoic acid (bp = 36 °C) or an inorganic azide, which has the potential to form hydrazoic acid *in situ*. Given its explosivity[20] and toxicity,[21] it is prudent to avoid conditions that could lead to the formation of hydrazoic acid during scale-up.

Although this approach was effective to provide gram quantities of **1** to support early SAR and pre-clinical toxicology studies, there were several safety concerns regarding the use of sodium azide at elevated temperatures, especially on a large-scale.[21] This prompted our quest for a safer alternative that would be more amenable to scale-up.

7.2 Results and Discussion

We were intrigued by Dimroth's 1910 report on the synthesis of 1-aryl-1*H*-tertazoles (Scheme 7.3).[22] This approach involves the reaction of an aryl diazonium salt with diformylhydrazine to yield the desired tetrazole. The synthesis of tetrazoles was not the primary objective of the original work; nevertheless, since then very few reports of application of this methodology for the synthesis of tetrazoles have emerged.[23,24] At a cursory glance, this procedure appeared to have several advantages over those described in Scheme 7.1: (a) this reaction does not involve the use of azides; (b) the conditions preclude the formation of hydrazoic acid; and (c) the reaction is carried out under aqueous conditions under which the product directly crystallizes out from the reaction mixture, rendering this method particularly appealing for scale-up.

Our initial attempts with the Dimroth approach were met with limited success. While it was heartening to note that the reaction provided the desired product, the modest conversions achieved (*ca.* 30%) suggested that further exploration was required. We envisioned that reinvestigating this reaction, and understanding its scope and limitations would pave the way for the development of a safe, robust protocol for the multi-kilogram synthesis of the desired 1-aryltetrazole.

Horwitz *et al.* postulated that this reaction proceeded by the pathway depicted in Scheme 7.3.[24] Nucleophilic attack of the anion of

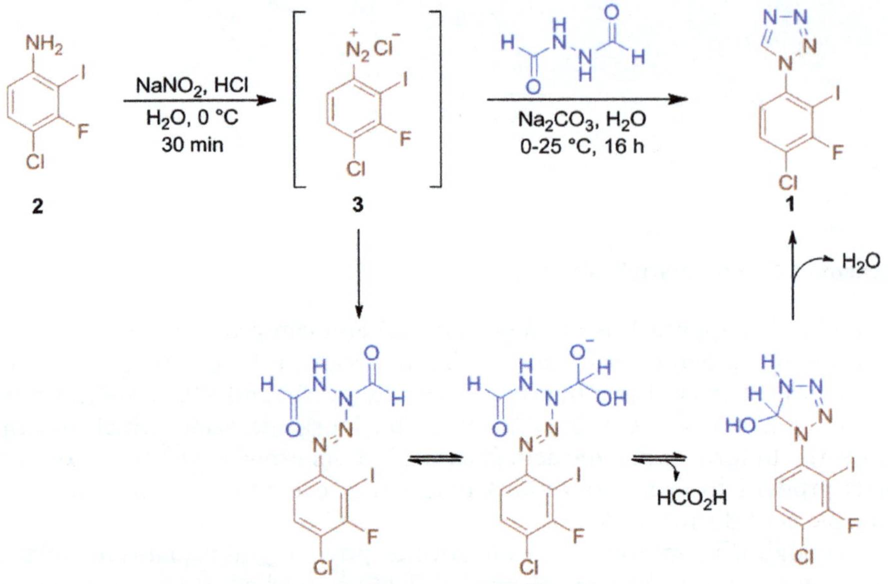

Scheme 7.3 Synthesis of 1-aryl-1*H*-tertazoles.

diformylhydrazine onto the diazonium salt would lead to tetrazene **4**. Attack of aqueous base on **4** would provide **5**, which upon loss of formic acid, cyclization and dehydration would furnish the desired tetrazole **1**.

We verified this hypothesis by isolating tetrazene **4** following Mackay's protocol;[25] subsequent treatment of this solid with NaOH afforded tetrazole **1**. It must be emphasized that tetrazene **4** was isolated primarily to garner insights into the mechanism of the reaction. Isolation of this intermediate was not viable on scale-up because of the material losses incurred during isolation, and more importantly, due to its reported instability.[25]

In this experiment, we observed that while the initial tetrazene formation occurred cleanly, the cyclization was sluggish while using NaOH as the base, leading to only 30% conversion. This led us to speculate that the cyclization itself was sensitive to the pH of the reaction, while the tetrazene formation would be dependent on the stoichiometry of diformylhydrazine.

We embarked on a systematic study varying the pH of the reaction and the equivalents of diformylhydrazine to understand their impact on the reaction profile. A two-factorial design using a quadratic model was explored, and the results are depicted in Figure 7.1. From the data, it was evident that the best conversions (~65%) were achieved when the pH of the system was maintained between 7 and 9 throughout the reaction with 2 equiv. of diformylhydrazine. A drastic drop in conversion was observed outside this pH range, emphasizing the need for strict pH control during this reaction.

Having established the optimum pH range and stoichiometry for executing this transformation, we focused our attention on the safety aspects of

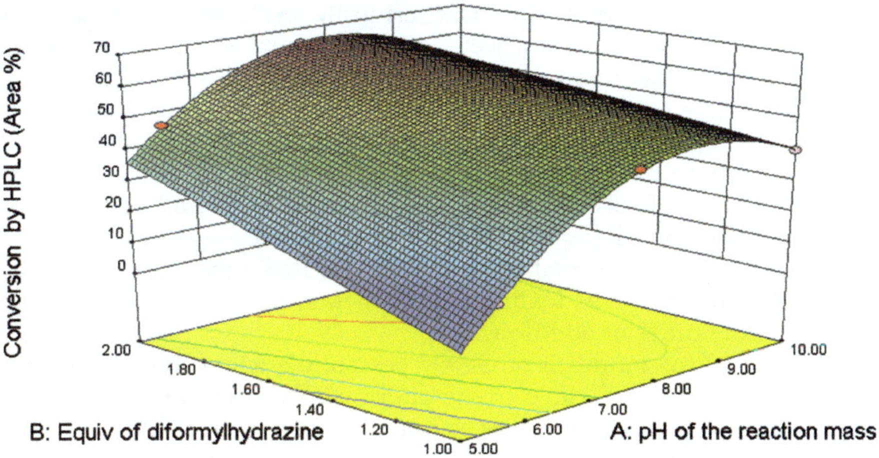

Figure 7.1 Effect of pH and equivalents of diformylhydrazine on reaction profile.

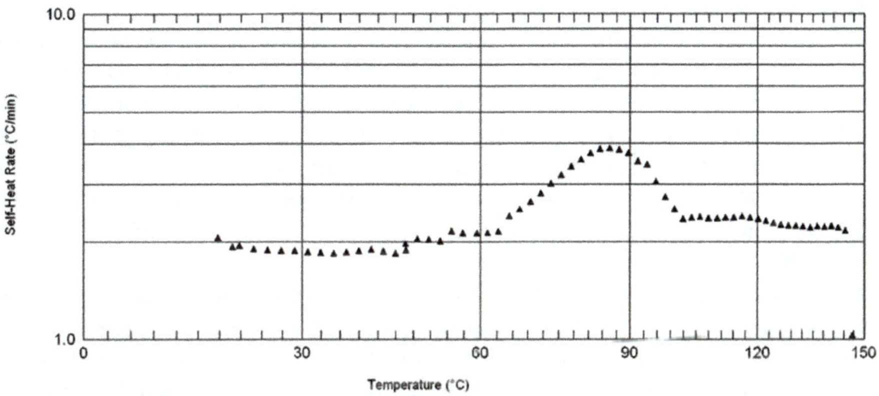

Figure 7.2 ARSST plot of reaction mixture containing diazonium salt **3**.

this reaction to ensure scalability. Diazotization reactions are known to be exothermic, and several diazonium salts are known to be unstable at elevated temperature. Reaction calorimetry studies indicated that the formation of diazonium intermediate **3** (*via* addition of sodium nitrite to amine **2**) was mildly exothermic with an adiabatic temperature rise of 5 °C. After the completion of reaction, the thermal stability of the diazonium salt-containing reaction mixture was analyzed by differential scanning calorimetry (DSC) and Advanced Reaction System Screening Tool (ARSST). The reaction mixture showed an onset of exotherm/self-heating at ~45 °C (Figure 7.2) with a minor energy change of −60 to −70 J g^{-1}, and an adiabatic temperature rise (ΔT_{ad}) of 15–20 °C. Hence it was decided to maintain the reaction temperature below 10 °C to avoid decomposition of the diazonium intermediate.

The tetrazole formation step was critically evaluated from a safety perspective. In an effort to avoid accumulation of tetrazene intermediate **4**, it was decided to add a pre-mixed solution of diformylhydrazine and aqueous sodium carbonate to the diazonium reaction mass. This addition was found to be mildly exothermic ($\Delta T_{ad} \sim 2$ °C), but vigorous foaming (doubling of reaction volume from 250 mL to 500 mL) was observed during the addition due the evolution of carbon dioxide on a 17 g scale. Around 180 L of CO_2 gas was evolved per kilogram of starting material, and a peak flow rate of 10 L min^{-1} kg^{-1} was recorded by the Ritter gas flow meter connected to a 500 mL RC1e reactor. This foaming phenomenon prompted us to consider redesigning the order of addition.

The foaming was found to be barely perceptible when the intermediate diazonium solution was added to the diformylhydrazine-sodium carbonate solution, and the gas evolution was also significantly reduced (30 L kg^{-1} with respect to **2**), with no major spike in gas flow rate during the course of addition. This reverse addition protocol circumvented the foaming problem that was encountered in the previous approach, and also ensured that the reaction mixture (for tetrazole formation) was basic at all times, leading to optimal conversion.

The reaction mass after the formation of tetrazole **1** was heterogeneous, and hence the thermal stability of the wet solid and the mother liquor were separately assessed by DSC. The aqueous layer showed onset of minor exotherm around 110 °C and 190 °C with an adiabatic temperature rise of 3–5 °C. The crude tetrazole product showed an onset of minor exotherm around 74–115 °C (Figure 7.4B) with energy of −46 J g^{-1} ($\Delta T_{ad} = 25$ °C) and onset of major exotherm around 116–217 °C with energy of −518 J g^{-1} ($\Delta T_{ad} = 300$ °C).

The minor exotherm observed in the crude tetrazole at 74 °C was too close to the intended drying temperature of 40–45 °C, and hence warranted further investigation. LCMS analysis of the crude product revealed the presence of 1-chloro-2-fluoro-3-iodobenzene **7** and cyanamide **8** (Figure 7.3) as minor impurities along with a small quantity of unreacted diazonium salt **3**. An aqueous NaOH wash of the crude product removed these impurities as confirmed by DSC and/or LCMS, and consequently eliminated the exotherm at 80 °C as evidenced by DSC (Figure 7.4C).

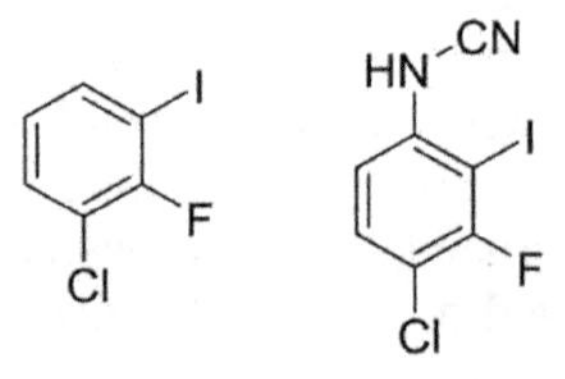

7 8

Figure 7.3 Structures of impurities in the crude tetrazole.

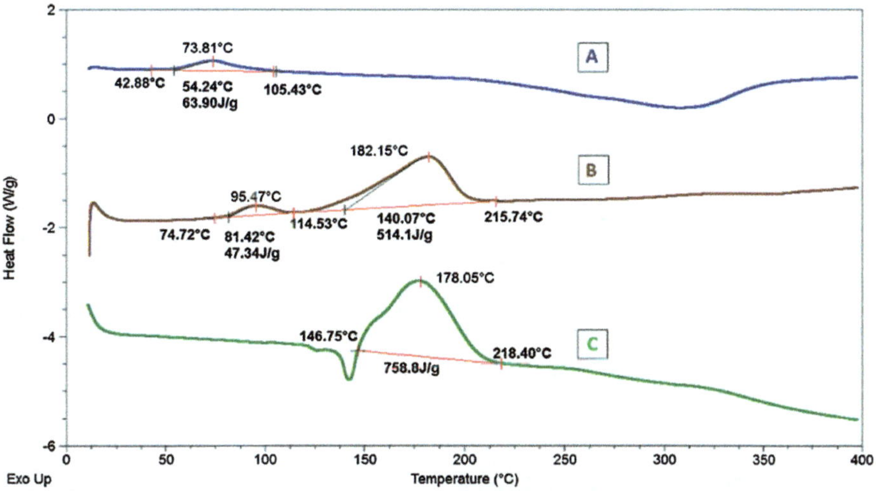

Figure 7.4 DSC thermograms of the intermediate diazonium reaction mass (A), crude tetrazole solid (B) and isolated tetrazole (C).

Since the final tetrazole product was to be dried at 40–45 °C, it was also important to assess the thermal stability of tetrazole **1** at elevated temperature. The DSC scan of isolated tetrazole **1** (Figure 7.4) showed a major exotherm around 146 °C with energy of −760 J g^{-1} ($\Delta T_{ad} = 448$ °C) The heat data obtained from dynamic runs (scanning rate 4, 6, 8, 10 and 12 C min^{-1}) in the DSC were exported to the AKTS software to predict the adiabatic decomposition temperature for 24 h (ADT 24).[26] AKTS software using isoconversional methods predicted the maximum safe drying temperature for the tetrazole product to be 93 °C (Figure 7.5) based on ADT 24. This was found to be significantly above the intended drying temperature of 40–45 °C.

7.3 Conclusion

In conclusion, we have developed an improved, efficient, and safe process for the synthesis of tetrazole **1** under aqueous conditions from diazonium salts, which offers distinctive advantages over conventional procedures. Specifically, this protocol obviates the need to use sodium azide, but instead uses commercially available diformylhydrazine, which is considered much safer and amenable to scale-up. A systematic study of the reaction parameters led to the definition of the optimum pH range and stoichiometry to effect this transformation. The order of addition of the reagents and starting materials was designed to minimize major exotherms and gas evolution. This process has been successfully utilized for the synthesis of 1-(4-chloro-2-iodophenyl)-1*H*-tetrazole on a multi-kilogram scale. We believe that this

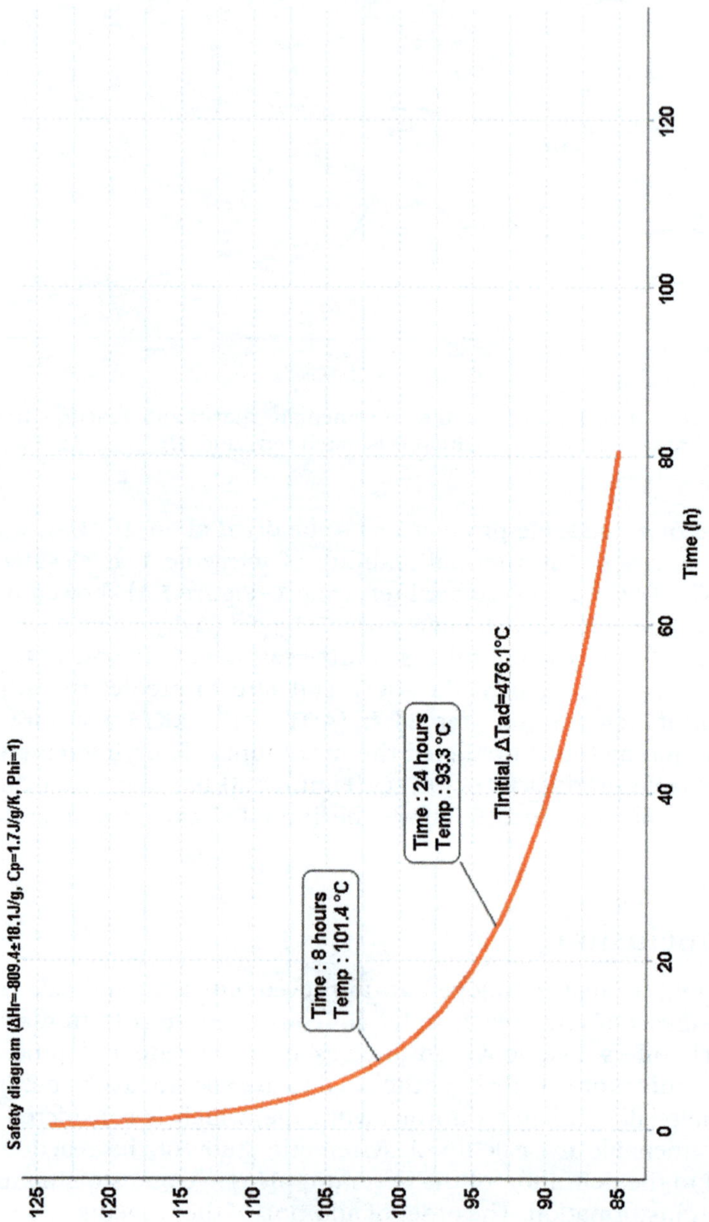

Figure 7.5 Thermal safety diagram for isolated tetrazole 1.

approach is general and can be extended to the synthesis of a host of 1-aryltetrazoles, and should therefore find widespread use.

7.4 Representative Procedure

1-(4-Chloro-3-fluoro-2-iodophenyl)-1H-tetrazole (1): To a suspension of 4-chloro-3-fluoro-2-iodoaniline hydrochloride (7.6 kg, 24.68 mol) in 42 L of water 15.2 L of concentrated hydrochloric acid was added. A solution of sodium nitrite (2.2 kg, 31.88 mol) in water (30 L) was added dropwise with stirring at 0–3 °C. The reaction mass was stirred for 1 h at 0–3 °C to get a clear yellow coloured solution containing diazonium salt **3**. In a separate reactor, diformyllhydrazine (4.4 kg, 49.96 mol) and 230 L of 10% aqueous sodium carbonate solution were taken and cooled to 0–5 °C. The diazonium salt solution was added to the diformylhydrazine-sodium carbonate solution, maintaining the temperature at 0–10 °C. The mixture was stirred overnight at 20–30 °C, and the resulting solid was collected by filtration. The wet cake was transferred to a reactor, and stirred with 38 L of 10% aqueous sodium hydroxide solution at room temperature for 2 h, and filtered. The filter cake was washed with 45 L of water and the crude brown solid was slurried with a 2 : 1 mixture of MTBE and *n*-heptane for 30 min at ambient temperature. The resulting solid was washed with 25 L of *n*-heptane, and dried *in vacuo* at 40–45 °C for 10 h to provide 5.3 kg (65% yield) of 1-(4-chloro-3-fluoro-2-iodophenyl)-1*H*-tetrazole **1** as a pale brown solid. ^{1}H NMR (DMSO-d_6) δ 9.90 (s, 1 H), 7.91–7.96 (m, 1 H), 7.64 (dd, $J = 8.4$, 1.5 Hz, 1 H); ^{13}C NMR (DMSO-d_6) δ 157.3 (d, $J_{C,F} = 243$ Hz), 144.8, 136.74 (d, $J_{C,F} = 3$ Hz), 131.2, 124.9 (d, $J_{C,F} = 3.8$ Hz,), 122.3 (d, $J_{C,F} = 21.1$ Hz), 87.9 (d, $J_{C,F} = 29.4$ Hz). Anal. Calcd for $C_7H_3ClFIN_4$: Elemental Analysis: C 25.91, H 0.93, N 17.27; Found: C, 25.74, H 0.99, N 16.85%.

Acknowledgements

The authors thank Rajendra Deshpande, Nicolas Cuniere, Monica Fitzgerald, Mujeebur Rahuman, Yogitha Murthy, and Loganathan Chandrasekaran for helpful discussions during the course of this work. Somprabha Sidar's help in acquiring process safety data is gratefully acknowledged. We also thank Nachiket Likhite, Rajappa Vaidyanathan, Prantik Maity and Martin Eastgate for their helpful suggestions on this manuscript, and David Kronenthal for support of this work.

References

1. R. J. Herr, *J. Bioorg. Med. Chem.*, 2002, **10**, 3379; R. N. Butler, in *Comprehensive Heterocyclic Chemistry II*, ed. A. R. Katritzky, C. W. Rees and E. F. V. Scriven, Pergamon, Oxford, UK, 1996, vol. 4, p. 621; L. J. Kennedy, *Tetrahedron Lett.*, 2010, **51**, 2010; R. R. Wexler, W. J. Greenlee, J. D. Irvin, M. R. Goldberg, K. Prendergast, R. D. Smith and P. B. M. W.

M. Timmermans, *J. Med. Chem.*, 1996, **39**, 625; K. Noda, Y. Saad, A. Kinoshita, T. P. Boyle, R. M. Graham, A. Husain and S. S. Karnik, *J. Biol. Chem.*, 1995, **270**, 2284.
2. J. A. Bladin, *Ber. Dtsch. Chem. Ges.*, 1885, **18**, 1544.
3. R. J. DeVita and M. J. Wyvratt, *Drugs Future*, 1996, **21**, 273.
4. J. S. Morley, *J. Chem. Soc. C*, 1969, 809; C. W. Thornber, *Chem. Soc. Rev.*, 1979, **8**, 563; H. Singh, A. S. Chawla, V. K. Kapoor, D. Paul and R. K. Malhotra, *Prog. Med. Chem.*, 1980, **17**, 151; R. N. Butler, *Adv. Het. Chem.*, 1977, **21**, 323; S. J. Wittenberger, *Org. Perp. Proced. Int.*, 1994, **26**, 499; A. Burger, *Prog. Drug Res.*, 1991, **37**, 287; C. A. Lipinski, *Annu. Rep. Med. Chem.*, 1986, **27**, 283.
5. G. F. Holland and J. N. Pereira, *J. Med. Chem.*, 1967, **10**, 149; S. K. Figdor and M. Schach von Wittenau, *J. Med.Chem.*, 1967, **10**, 1158; D. W. Esplin and D. M. Woodbury, *J. Pharmacol. Exp. Ther.*, 1956, **118**, 129; T. Kimura, D. Shuto, Y. Hamada, N. Igawa, S. Kasai, P. Liu, K. Hidaka, T. Hamada, Y. Hayashi and Y. Kiso, *Bioorg. Med. Chem. Lett.*, 2005, **15**, 211.
6. C. Hansch and L. Leo, *Exploring QSAR. Fundamentals and Applications in Chemistry and Biology*, American Chemical Society, Washington DC, 1995, ch. 13.
7. G. F. Holland and J. N. Pereira, *J. Med. Chem.*, 1967, **10**, 149; S. K. Figdor and M. Schach von Wittenau, *J. Med. Chem.*, 1967, **10**, 1158.
8. J. L. Kraus, *Pharmacol. Res. Commun.*, 1983, **15**, 183.
9. K. Kubo, Y. Kohara, Y. Yoshimura, Y. Inada, Y. Shibouta, Y. Furukawa, T. Kato, K. Nishikawa and T. Naka, *J. Med. Chem.*, 1993, **36**, 2343; J. Alexander, M. L. Renyer and G. S. Rork, *J. Pharm. Sci.*, 1994, **83**, 893; D. E. Ryono, J. Lloyd, M. A. Poss, J. E. Bird, J. Buote, S. Chong, T. Dejneka, K. E. J. Dickinson, Z. Gu, P. Mathers, S. A. Moreland, R. A. Morrison, E. W. Petrillo, J. R. Powell, T. Schaeffer, E. R. Spitzmiller and R. E. White, *Bioorg. Med. Chem. Lett.*, 1994, **4**, 201.
10. J. Sauer, R. Huisgen and H. J. Strum, *Tetrahedron*, 1960, **11**, 241.
11. A. I. Lesnikovich, O. A. Ivashkevich, S. V. Levchik, A. I. Balabanovich, P. N. Gaponik and A. A. Kulak, *Thermochim. Acta*, 2002, **388**, 233.
12. M. Mihit, R. Salghi, S. El Issami, L. Bazzi, B. Hammouti, E. A. Addi and S. Kertit, *Pigment Resin Technol.*, 2006, **35**, 151.
13. T. Jin, S. Kamijo and Y. Yamamoto, *Tetrahedron Lett.*, 2004, **45**, 9435.
14. F. G. Fallon and R. M. Herbst, *Chem. Rev.*, 1947, **41**, 1.
15. E. O. Mandalà and B. Alagna, *Gazz. Chim. Ital.*, 1910, **40**, 441.
16. M. Freund and T. Paradies, *Ber. Dtsch. Chem Ges.*, 1901, **34**, 3110; R. Stollé and F. Henke-Stark, *J. Praki. Chem.*, 1930, **124**, 261.
17. J. von Braun and W. Rudolph, *Chem. Ber.*, 1941, **74**, 264.
18. Y. Satoh and N. Marcopulos, *Tetrahedron Lett.*, 1995, **3**, 1759; A. K. Gupta, C. H. Song and C. H. Oh, *Tetrahedron Lett.*, 2004, **45**, 4113.
19. W. Su, Z. Hong, W. Shan and X. Zhang, *Eur. J. Org. Chem.*, 2006, 2723.
20. P. G. Urben, *Bretherick's Handbook of Reactive Chemical Hazards*, Elsevier, 7th edn, 2007, vol. 2, p. 96.

21. F. González-Bobes, N. Kopp, L. Li, J. Deerberg, P. Sharma, S. Leung, M. Davies, J. Bush, J. Hamm and M. Hrytsak, *Org. Process. Res. Dev.*, 2012, **16**, 2051.
22. O. Dimroth and G. de Montmollin, *Ber. Dtsch. Chem. Ges.*, 1910, **43**, 2904.
23. D. Wu and R. M. Herbst, *J. Org. Chem.*, 1952, **17**, 1216.
24. J. P. Horwitz and V. A. Grakauskas, *J. Org. Chem.*, 1954, **19**, 194; J. P. Horwitz and V. A. Grakauskas, *J. Am. Chem. Soc.*, 1957, **79**, 1249.
25. D. Mackay and D. D. McIntyre, *Can. J. Chem.*, 1982, **60**, 990.
26. T. Lakshminarasimhan, *Org. Process. Res. Dev.*, 2014, **18**, 315.

CHAPTER 8

New Directions from Academia

MANAVI YADAV, SRIPARNA DUTTA AND
RAKESH KUMAR SHARMA*

Green Chemistry Network Centre, Department of Chemistry,
University of Delhi, Delhi-110007, India
*Email: rksharmagreenchem@hotmail.com

8.1 Introduction

In the modern era of globalization, academia–industry collaborations have become increasingly significant on account of growing economic pressure, fierce competition from emerging economies and the rising perplexity of the problems.[1] Today, the pharmaceutical industries (known for their incredible discoveries in the form of life-saving drugs) are afflicted with some serious problems such as declining research and development (R&D), low productivity, share losses and reduced profit potential.[2] Faced with low records in drug approval owing to late-stage failures, lack of innovation in the research pipeline and the probability of the blockbuster drugs to go off-patent in the next few years, the pharmaceutical enterprises have started to realize that teaming up with academic institutions can not only breathe new life into translational research, providing them the right direction while decreasing the R&D budget, but also help tackle the challenges of drug discovery and development.[3] While working synchronously with the academicians, industrial researchers are exposed to the most advanced level of creative as well as technical thinking, new experimental strategies and breakthrough research trends. Besides, the acting practitioners *i.e.* industrialists gain discussion partners who can actually question their beliefs and conclusions, thereby providing them an opportunity to validate their directions. Such

Green Chemistry Series No. 52
Hazardous Reagent Substitution: A Pharmaceutical Perspective
Edited by Rakesh Kumar Sharma and Rakeshwar Bandichhor

Published by the Royal Society of Chemistry, www.rsc.org

close collaborations also prove to be advantageous for the reflecting researchers *i.e.* the academicians as they get a chance to translate their basic discoveries into new therapeutics and simultaneously get access to resources and funding during times of dwindling grant support (Figure 8.1).[4,5]

In fact, these two domains, namely academia and industry, share a very rich history of productive collaboration dating back to the early 20th century. Amongst the earliest examples that illustrate the victories of academic–industry alliances were the discovery of two crucial life-saving respiratory drugs: **corticosteroids** that won a Nobel Prize for Philip Hench (Mayo Clinic), Tadeous Reichstein (Basel University) and Edward Kendall (Mayo Clinic); and **streptomycin** (discovered with the joint efforts of Selman Waksman, Rutgers University and Geoffrey Marshall, Medical Research Council, United Kingdom).[6,7] What is worth mentioning here is that both discoveries would not have been possible without the help of academic intelligence. The striking success of corticosteroids and streptomycin ignited the research interest of the scientific community to unravel and exploit the vast molecular diversity of nature in search of novel molecular architectures, biological tools and lead compounds for the drug discovery process. Consequently, as a result of the devoted efforts of academic and industrial researchers, several other drugs could be discovered to fight deadly diseases such as human immunodeficiency virus (HIV), cancer, *etc.* There is another instance which brilliantly highlights the role of the academia in bringing a major reformation in the pharmacological sector. In a quest to exploit the newly discovered recombinant DNA technologies from the University of California at Berkeley, Stanford and the University of California at San Francisco, Genentech—a biotech company—was set up in the year 1976.[8,9] Outstandingly, this company has become a pioneer in the area of therapeutic proteins as it has been able to produce exceptional products, some of which include human insulin (the first medicine to be approved by the FDA made from rDNA technology, licensed to Eli Lilly), recombinant growth hormones (such as protopin) and pulmozyme (for the treatment of cystic fibrosis). By fostering an innovative internal research program that rivals that of many academic institutions, Genentech has managed to stay at the forefront of

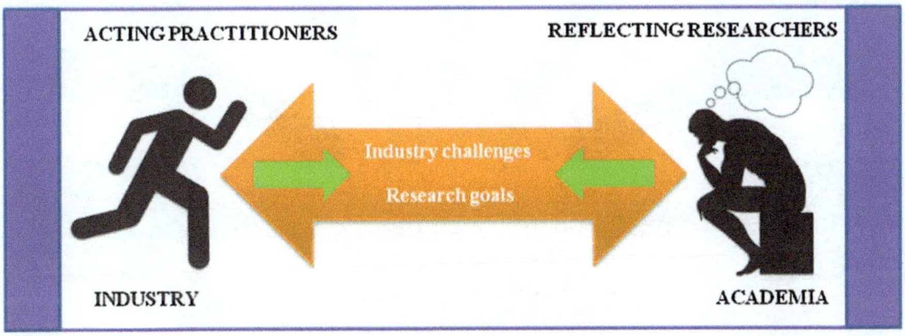

Figure 8.1 Academia and industry working together.

protein therapeutic development. The success of such close partnerships has aroused the interest of federal and state policy makers in expanding the role of universities from purely academic training centres to institutions that can drive economic development and innovation, thus changing the historical notion of perceiving research institutes as a source of new ideas and industries as the route for maximising the use of these ideas.

Recent years have witnessed a noticeable resurgence of partnering that is poised to address many of the modern challenges for advancing research, innovation and technological development. However, the impact of technology expansion has raised serious concerns on the global environment. Amongst other industries, the chemical industry is considered the biggest source of toxic waste and this is continuously rising due to a series of more restrictive laws. These wastes can be process residues, spent catalysts, solvent effluents, treatment sledges and/or contaminated chemical containers that have a detrimental impact on the environment. Not only this, but the reagents employed in various organic transformations that afford important moieties for synthesizing APIs and pharmaceutical intermediates are considered to be hazardous due to the formation of highly reactive and exothermic intermediates, and are therefore not suitable for implementation on a large scale. For these reasons, there is a growing interest in the new frontiers of Green Chemistry to develop alternative approaches that enable these transformations to be performed safely on a commercial scale.[10] Several strategies that have been developed to minimize waste generation have been summarized in Figure 8.2.

The ensuing sections shall throw light upon: (i) Green Chemistry: a functional approach winning industrial support; (ii) the need for academia–industry collaborations in the current scenario; (iii) current

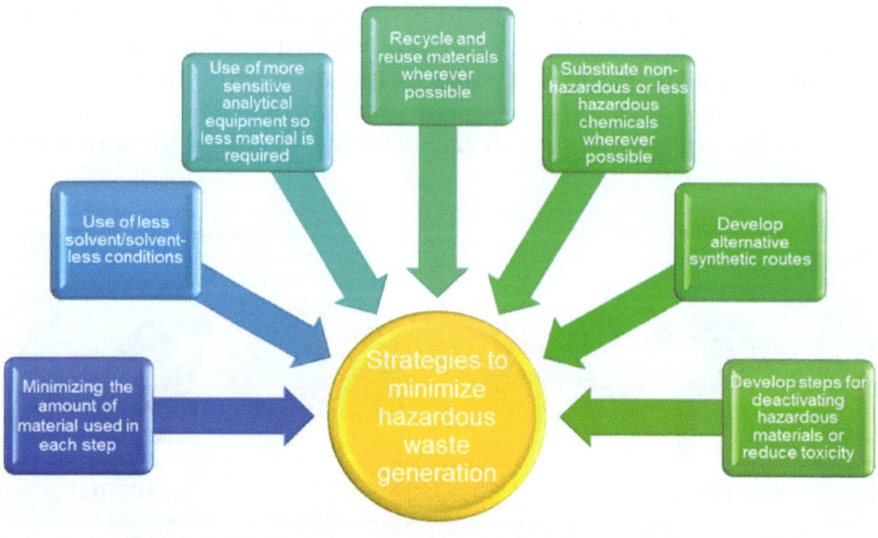

Figure 8.2 Strategies to minimize hazardous waste generation.

obstacles to academic–industrial collaboration; (iv) keys for overcoming these obstacles (new models for collaboration); (v) greening of industries: promoting newer and greener methodologies from academia; (v) benefits of greening industries: industrial case studies; (vi) cases of innovations/novel technologies stemming from successful partnerships between these two sectors (case studies where academia provided new directions to industry); and (vii) redefining the operational methods of global chemical industries towards sustainability.

8.2 Green Chemistry: A Functional Approach Winning Industrial Support

The term green chemistry was coined in 1990s, and has been credited to the design of chemical products and processes that reduce or eliminate the use and/or generation of hazardous substances. It has also spawned wide areas of research that include green solvents, alternative energy science, bio-based transformations and materials, molecular self-assembly, and next-generation catalyst design. It is an overarching approach that applies to all aspects of chemistry and permeates every stage of process development. In fact, there are reports that predict green chemistry as the future of all chemistry and it is predicted that the global market of green chemistry will grow exponentially to $98.5 billion by 2020.[11] At present, green chemistry and sustainability aim to address a number of challenges that have been outlined in Figure 8.3.

Despite significant investment over the past several years to ensure that further chemical production does not have malign impacts on the environment and future generations, there has been a continuous increase in the use of hazardous chemicals. This gave researchers a driving force towards developing green chemistry. From feed-stocks to product synthesis and processing, various attempts have been made to produce chemicals that are sustainable and benign to the environment and human health. Nowadays, the immediate challenge is to prevent pollution at its source rather than treating the waste. With the "command and control" approach, green chemistry also seeks to improve present technologies in greener and economical manner. Green chemistry has wide applicability in different sectors as well as the pharmaceutical industry (Figure 8.4).

8.3 Need for Academia–Industry Collaborations in the Current Scenario

Ever since the passage of the **US Bayh–Dole Act** or **Patent and Trademark Law Amendments Act** of 1980 which enabled universities to retain ownership of patents generated from federally-funded research programmes, thereby providing a breakthrough policy change in commercialization of knowledge and innovations, the number of research partnerships between

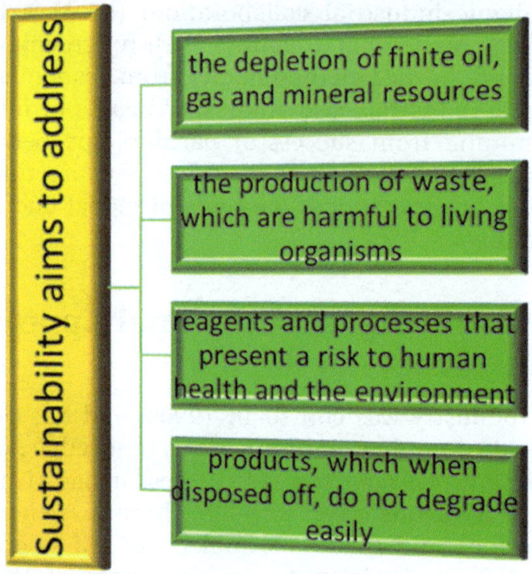

Figure 8.3 Aspirations for sustainability.

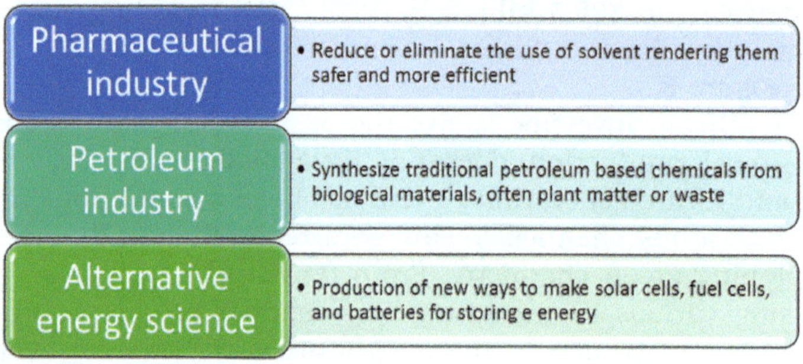

Figure 8.4 The wide range of applications of green chemistry.

industries and universities has risen exponentially.[12] Many trends have fuelled these alliances further and created an environment that fosters a co-dependency of these two sectors despite having substantive differences have been identified and listed as:

Dependency of industry on academia: The industrial sector relies on academic scientists for basic research that identifies novel molecular targets and evaluates the efficacy and safety of inventions from industry.

Identified causes for increased dependency in current times are:

- *Declining research and development productivity*
- *Expiry of key drug patents*
- *Loss of shareholder value*

- *Reduced profit potential*
- *Increasing cost of bringing new molecular entity to market*

Dependency of academia on industry: Academicians need industrial support to translate their basic discoveries into new therapeutics and to procure medicines and technologies that are required for taking care of patients.

Identified causes for increased dependency in current times are:

- *Decrease in government funding of academic research*
- *Decrease in academic jobs – the number of available jobs is vastly outnumbered by the number of people applying for them*

Reflecting on the points mentioned above, it becomes quite clear that there is not only a natural complementarity between academia and industry that has resulted in the increased collaborations; but collaborations have become exceedingly important in view of the problems faced by each of these domains.[13,14]

8.4 Impediments to Collaboration

Although pharma-academia research-related relationships have been understood to be essential in the discovery of innovative therapies and revolutionary drugs, unfortunately the interface between them has been a source of tension in the past and continues to be highly problematic even now.[15–17] It is important for us to understand the factors responsible for the potential pitfalls (*i.e.* behind the failing relationships) and the challenges that need to be overcome for successful partnerships. Amongst the identifiable causes, the differences in the core objectives of the industrialists and scientists and the way they work shows why there is an unwillingness to work together despite knowing the benefits of such close collaborations. In addition, the policies concerning collaboration, mutual lack of understanding expectations, the time and cost involved in clinical trials of the compounds *etc.* again discourage them from engaging in partnerships. We have elaborated these impediments below:

A. **The mentality gap and difference in work culture between industry and academia**
   - While universities work with the prime goal of educating the masses, creating new knowledge, and excelling in the existing know-how, the pharmaceutical companies strive to work for market success by mastering the challenges of a competitive environment.
   - The success of academic research is measured by the number of articles published and the impact it will have on the global research community. Research carried out in industry, on the other hand, is significantly driven by the impact it will have on the company. As a result of varying interests and work cultures, these parties differ drastically in their perceptions and judgements about

organizational structures, decision-making processes and time scales which lead to conflicts and disagreements.

B. **Policy concerning collaborations:** Every academic organization has its own policy related to collaboration with industry. The lack of availability of a widely adopted template greatly hampers efficiency and speedy research.
C. **Irreproducibility of data from academia:** One of the most frequently encountered problems faced by industries while working with academicians is irreproducibility of data. On evaluating 70 targets, scientists at Bayer found that for almost two-thirds of the targets, *i.e.* their initial basic research data that had generated interest, could not be replicated. This indeed is a serious problem.
D. **Delay in publication:** Because firms are interested in obtaining new patents or new products, they want to delay publications to avoid disclosing information; university researchers whose prime motive is to publish results as fast as possible refrain from engaging in such relationships.
E. **A prolonged and expensive process:** Collaborations involve a lot of time and investment, however at times when no fruitful results are obtained; it gets highly frustrating for the firms as they seek short-term results and clear contributions to the current business lines.
F. **Difficulties in negotiating collaborations:** Factors such as lack of adequate information, problems encountered in finding the right contact person, the transaction costs of finding the right partner *etc.* create difficulty in negotiating collaborations.
G. **Red Tape – a barrier:** A strong clinical research component is required for any successful translational research venture; however. training clinical investigators requires a fair amount of time and money and around 50% of them drop out during the first trial, as a consequence of which first time investigators remain the last time investigators. Therefore, clinical research has become scarce in quite a few parts of the world.

8.5 Keys to a Successful Collaboration

So, the most important question of the hour is how to forge a higher level of strategic partnership. As rightly pointed out by Jăck Pronk, professor at Delft University of Technology *(stated in the March issue of Nature Biotechnology)*, the most important way to make academia–industry partnerships work is a willingness of the collaborators to understand and respect each other's core objectives (since the mentality gap is the most crucial issue that needs to be tackled).[18–20] Almost half the problems and challenges are resolved if a mutual respect is developed and this can be done if initiatives are taken by industries to invite academicians to interact with the practitioners at their facility. But this should be done without defining any outcome or result of the visit. The idea is the concerned people from both sides start spending

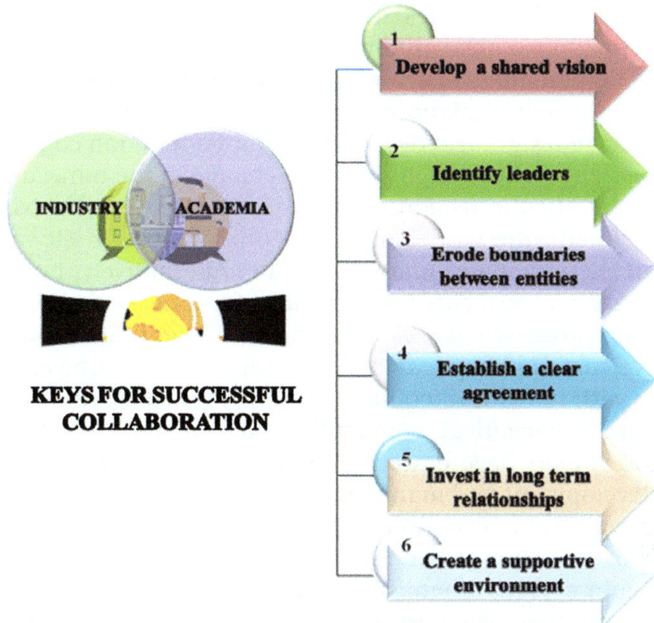

Figure 8.5 Keys for successful industry–academia collaboration.

reasonable time together so that their research interaction is given impetus. Apart from simply discussing problems, it is also important to involve the visitor in exploratory projects. It is anticipated that these visits can also help the academicians as they can make their research more meaningful and can also appreciate industry problems. To provide a holistic view to the readers, we have designed a key to successful industry–academia partnerships (Figure 8.5).

8.6 Greening of Industries: Promoting Newer and Greener Methodologies from Academia

The concept of green industry is to promote sustainable processes both in production and consumption of goods by means of using resources and energy efficiently, phasing out toxic substances, substituting fossil fuels with renewable energy sources, improving occupational health and safety, and reducing the overall risks.[21] Moreover, the final products must be such that can be easily managed throughout their life cycle. With this agenda, industries continuously work to improve processes and products along with their ecological performance. In fact, the greening of industries has become a principal element for fiscal competitiveness and sustainable growth. This is because resource inputs embody chief production cost and improve efficiency for industries, thereby giving them a competitive advantage. Not only this, but greening of industries also plays a vital role in poverty alleviation, growing occupations, and minimizing costs by increasing productivity. However, while greening, pharmaceutical industries face entirely different challenge in terms of redesigning a drug to make it more readily degradable. This is because redesigning a drug might alter its function completely as the drug development process might lead to a change in the molecular structure upon which therapeutic response depends.[22] Moreover, the existing drugs have already fought the battles of therapeutic tweaks, toxicity, and clinical trials and all these demand years to execute. Alternatively, by reducing the number of steps and reagents and by using solventless conditions, the drug manufacturing processes can be made more efficient. Still, the major challenge to drive sustainability is to search for methods that meet institutional expectations with industrial requirements. From the past few decades, many well-known pharmaceutical companies like Amgen, Merck, Johnson & Johnson, and Abbott have embraced green chemistry in practice for drug discovery, development and manufacturing. The driving force for this shift is the competitive advantage delivered by cheaper and environmentally benign processes.[23] In fact, chemical manufacturers such as DuPont have dedicated staff who explore green chemistry and incorporate life cycle analysis to products and manufacturing processes. The European Union's regulation on Registration, Evaluation, Authorisation and Restriction of Chemicals legislation has been a very dominant promoter of green chemistry and sustainable development. It favours innovative materials and processes and facilitates the phasing out of hazardous chemicals and their substitution with safer alternatives.[24] Table 8.1 provides examples of substituting non-hazardous or less toxic materials in chemical processes and experiments.

Implementation of green chemistry principles in the pharmaceutical industries have resulted in the successful discovery of several blockbuster drugs with reduced global environmental footprints. Also, there has been a significant improvement in the drug manufacturing processes in terms of either reaction conditions or reduction in the usage of solvents/reagents.

Table 8.1 Examples of substituting non-hazardous or less toxic materials in chemical processes and experiments.

Hazardous chemicals	Usage	Substitute
Xylene, benzene and toluene containing reagents	Histology laboratories	Citric acid based reagents
Solvent based inks	Printing operations	Soy based inks
Formaldehyde	Cleaning hospital kidney dialysis machines	Per-acetic acid
Mercury	Thermometers	Non-mercury thermometers
Sulfuric acid/potassium dichromate (chromerge)	Cleaning solutions	Detergents and enzymatic cleaners
Phosphate chloride detergents	Lab glass washing	Non-phosphate, non-chloride detergents

Notably, Viagra (sildenafil citrate)—an important oral medication used to treat erectile dysfunction—is manufactured by Pfizer with a new strategy under a reduced amount of solvent, eliminating the use of environmental pollutants such as tin chloride and hydrogen peroxide (a fire and transportation hazard).[25,26] Besides this, Pfizer also improved the process of manufacturing the popular drug Lipitor (atorvastatin) for reducing blood cholesterol by using an enzyme that catalyses the reaction in an aqueous medium.[27] Following the green chemistry footsteps, BASF also makes ibuprofen, a painkiller, in three steps rather than the conventional six step process, with a high yield.[28]

Working in green chemistry requires a unique set of expertise and resources that some companies are not well acquainted with/may still be developing. It is here where the role of academicians comes into play. There are examples where academic intervention has proven to be useful for these industries in either the development of a drug or the redesigning of existing processes. One such example includes the production of Zocor (simvastatin), drug to treat high cholesterol, which traditionally involved a multistep process and a large number of hazardous reagents. This has been replaced by a new method by Codexis (a biocatalysis company) that makes use of an engineered enzyme.[29,30] Another notable chemotherapy drug paclitaxel (marketed as Taxol) can be produced by growing tree cells in a fermentation vat, unlike the previous process that required a huge amount of solvent in addition to killing the tree.[31] To achieve sustainable processes leading to waste reduction, improvement in yield and safety, and eradicating the use of metal catalysts, as well as collaborating with academicians, industries are also partnering with each other. Examples include collaboration of Codexis with Merck for the development of a greener route to synthesize sitagliptin, the active ingredient in Januvia™ (a treatment for type 2 diabetes).[32]

8.7 Benefits of Greening Industries

Though environmental, human health and economic benefits all contribute as strong incentives for adopting green methodologies, economic benefits are considered to be the central drivers for industries in the advancement of greener technologies. The adoption of greener methodologies by industries improves corporate bottom lines by decreasing the wide array of operational costs. Environmental compliance automatically goes up when less waste is generated, and therefore, treatment and disposal of waste become unnecessary. Likewise, reduced solvent usage and fewer processing steps decrease the material and energy costs while manufacturing.

According to the United States Environmental Protection Agency (EPA) Toxic Release Inventory, the amount of chemical waste release to land, air and water decreased by 7% between 2004 and 2013. In fact, the decrease in the release of chemicals including hydrochloric acid, trichloroethylene, and methyl isobutyl ketone, was more than 60%. EPA analysis accredited this improvement to green chemistry and engineering practices. During the same time span, it was reported that the use of chemicals by U.S. drug industries dropped by nearly half as a result of using less organic solvents, such as methanol, dichloromethane, toluene, dimethylformamide, and acetonitrile.[33]

8.8 Innovations Stemming from Academia–Pharma Collaborations

8.8.1 Academic Contribution to Drug Discovery

For decades, scientific research leading to breakthrough innovations in the form of potent drugs, vaccines and other medical tools have enormously improved the quality of life, catalysing economic growth. According to the estimates of Pharmaceutical Research and Manufacturers of America, the death rate for cardiovascular diseases has fallen drastically, by almost 28%, with the development of prescription drugs while the average life expectancy for cancer patients has increased by three years since the 1980s. However, studies reveal that the big pharmaceutical companies have generally not been the fertile ground for innovation and have instead relied upon academia for fuelling their drug pipelines.[34,35] The dependence of the pharmaceutical industry on academia as the source of innovation can be readily judged from the analysis of the origin of all the European Medicines Agency approved drugs from 2010–2012. This analysis identified academia as a major source of innovation, thus highlighting the need for pharma to forge closer ties with universities.

8.8.1.1 Why Drug Discovery in Academia Makes Sense

Academic research has contributed immensely to advancing the frontiers of science and although there are several compelling reasons to justify why drug discovery in academia makes sense, we have listed only a few important ones which are as follows:

- Universities offer an excellent environment for creating new paradigms for drug discovery
- The diverse range of expertise available within universities facilitates the establishment of novel approaches in multidisciplinary areas like computational chemistry, synthetic organic chemistry, data mining, compound library selection and design, toxicity predictions *etc.*
- Academia is willing to take up projects of both discovery as well as methodology that involve high levels of risk and are not possible within the confinements of modern-day industries
- In contrast to the industrialists, academicians understand the value of individual freedom, intellectual diversity, flexibility and originality

The identification of a number of important drug candidates by academia that have eventually turned into blockbuster drugs have been enlisted in Table 8.2 which makes it clear that drug discovery is no longer the sole purview of industrial organizations.[36]

8.8.1.2 Case Studies

8.8.1.2.1 Darunavir (Purdue University & Tibotech/J&J). "Darunavir"—marketed under the brand name Prezista by Tibotec (a pharmaceutical and development company)—is a second-generation protease inhibitor used for treating HIV infection in adults and children above six years of age (Figure 8.6). Discovered by a medicinal chemistry professor Arun Ghosh and co-workers of Purdue University, *West Lafayette, US*, darunavir is recognized today as one of the most essential medicines by the World Health Organization (WHO).[37] Ghosh *et. al* used a structure-based design to combat drug resistance focusing primarily on the optimization of ligand–binding site interactions. The idea was to create an HIV protease inhibitor that would retain nearly full activity against multiple drug-resistant viruses, relative to its potency with wild-type viruses. To their delight, during investigation, they found that this drug has a superior activity against a panel of mutants that are resistant to all currently approved HIV protease inhibitors. Strikingly, this research has been enabled by the rich heritage of structural and chemical information available in the literature on HIV protease inhibitors and by productive collaborations with academic laboratories specializing in structural biology, virology and enzymology. In collaboration with Tibotec, the drug received accelerated FDA approval in June 2006 for use in combination with Norvir (ritonavir) for the treatment of HIV in adults that have previously experienced anti-retroviral treatment.

Mechanism of action:

The mode of action of darunavir is analogous to many of the other drugs utilized for treating retroviral infection. Once a cell is infected by HIV, it takes control of that cell, forcing it to make several copies of the virus. To make these copies, enzymes called proteases are used by the cell. This process of production of HIV slows down eventually when the activity of these enzymes is reduced. Darunavir blocks the activity of the HIV protease enzyme by forming robust interactions (*i.e.* hydrogen bonding) with the catalytic aspartates of its active site.[38] As a result of the strong binding of darunavir with the active site of protease, the formation of immature non-infectious viral particles occurs and the cells are no longer able to replicate further.

8.8.1.2.2 Emtriva (Emory University and Gilead Sciences). Another example of an academic contribution to HIV therapeutics is the discovery of **emtricitabine**, more commonly known as **Emtriva** – a nucleoside reverse

Table 8.2 Drugs with academic discovery origins.

Compound structure	Compound name	Target, indication	Academic institute	Industry partner
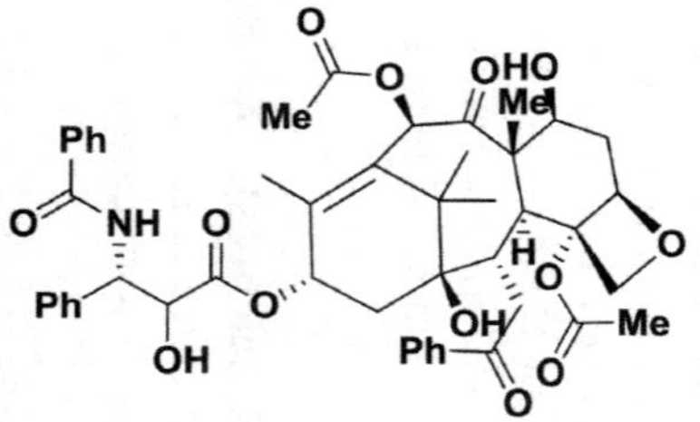	Paclitaxel, Taxol, Abraxane	Microtubule stabilizer, Oncology	Research Triangle Institute	Bristol-Myers Squib
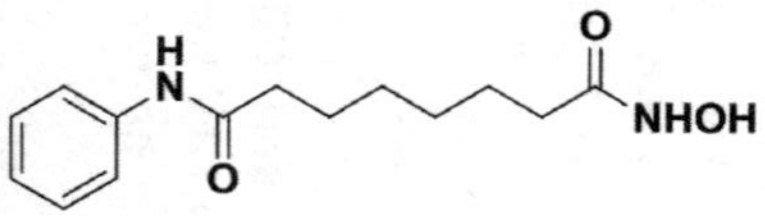	Vorinostat, SAHA	Histone deacetylase Inhibitor, Oncology	Sloan Kettering Cancer Center	Merck
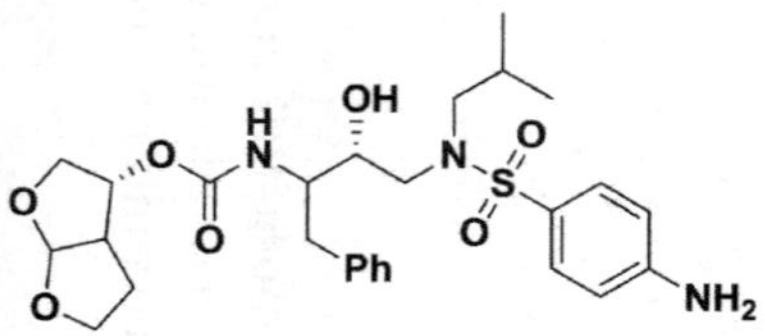	Prezista, Darunavir	HIV protease inhibitor, HIV	Purdue University	Tibotech/J&J
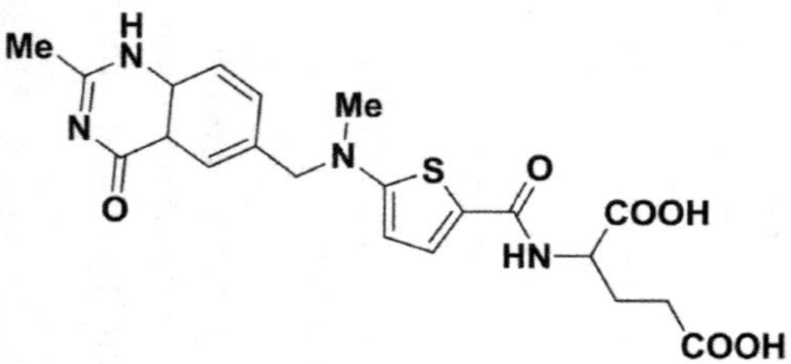	Tomudex, Raltitrexed	Thymidylate synthase inhibitor, Oncology	Institute of Cancer Research, UK	AstraZeneca

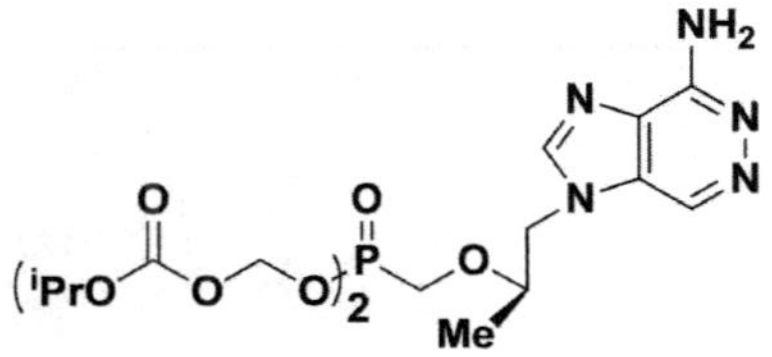	Viread, Tenovir disoproxil fumarate	Reverse transcriptase inhibitor, HIV infection	Institute of Organic Chemistry and Biochemistry Czech Republic, Rega Institute, Begium	Gilead
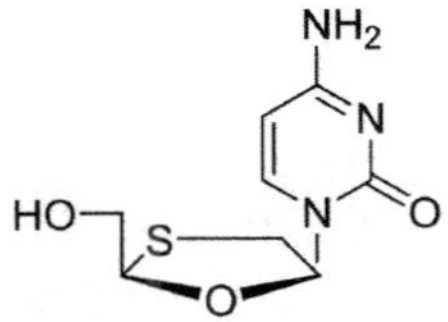	Lamivudine, (−)-3TC, Epivir	Reverse transcriptase inhibitor, HIV infection	Emory University	GSK
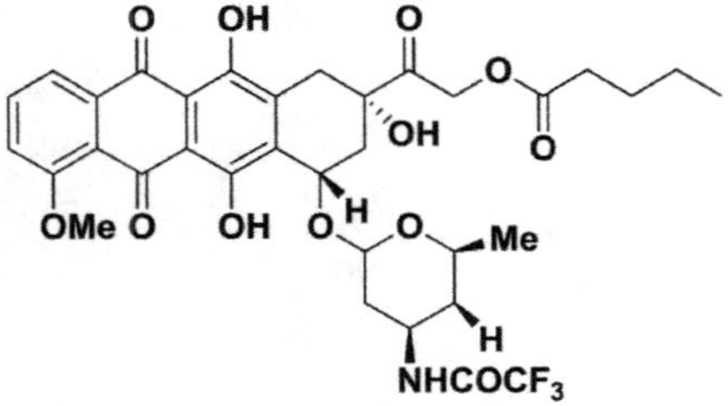	Valstar, Valrubicin	DNA topoisomerase II inhibitor, Oncology	Dana-Farber Cancer Institute, US	Endo Pharmaceuticals
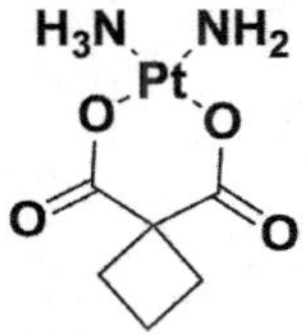	Paraplatin, Carboplatin	DNA alkylating agent, Oncology	Institute of Cancer Research, UK	Bristol-Myers Squib

Table 8.2 *(Continued)*

Compound structure	Compound name	Target, indication	Academic institute	Industry partner
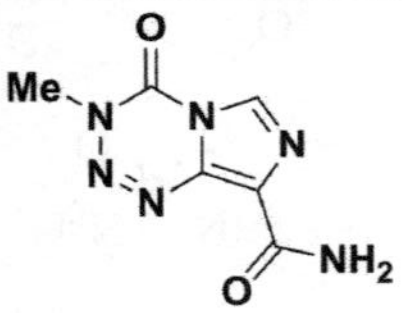	Temodar or Temodal, Temozolomide	DNA methylation, Oncology	Malcolm Stevens Aston University, UK	ScheringPlough Corp.
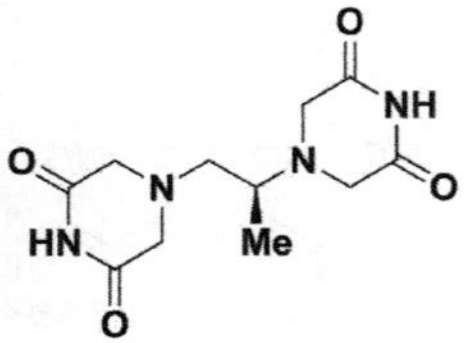	Dexrazoxane hydrochloride (Zinecard [Pfizer]; Cardioxane [Novartis])	Cardioprotective agent, Oncology	Cancer Research UK	Pfizer for USA & Canada Novartis for EU & ROW
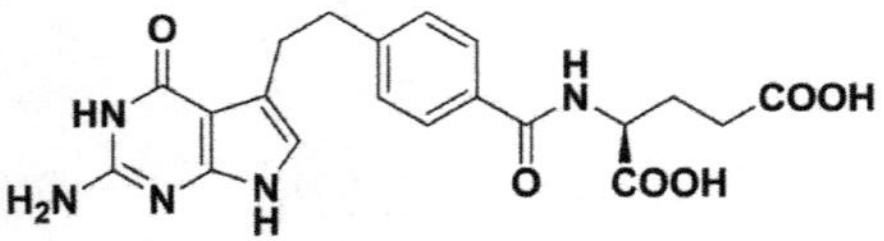	Alimta, Pemetrexed	Dihydrofolate reductase/ thymidylate synthase inhibitor, Oncology	Princeton University	Eli Lilly
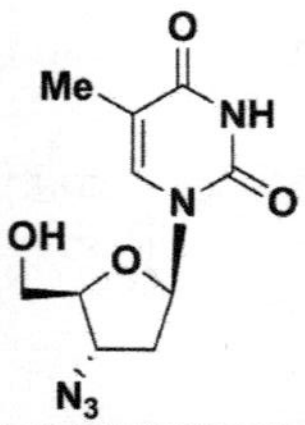	Zidovudine, AZT, Azidothymidine	Reverse transcriptase inhibitor, HIV	Michigan Cancer Foundation	GSK

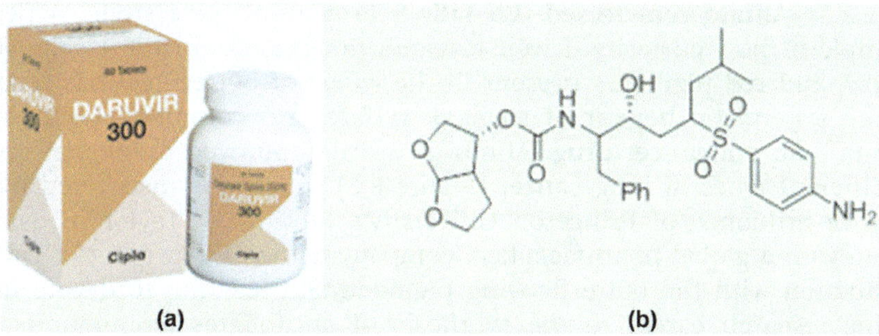

Figure 8.6 (a) Darunavir 300 mg tablet is used in the treatment of HIV infections and (b) Chemical structure of darunavir.

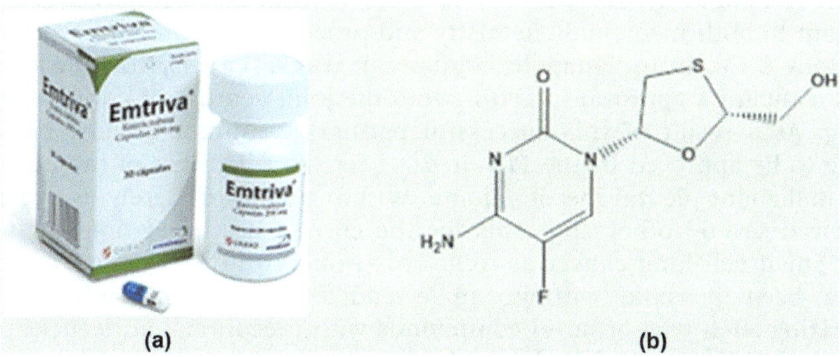

Figure 8.7 (a) Emtriva capsules 200 mg and (b) Chemical structure of Emtriva.

transcriptase inhibitor (NRTI) used for treating HIV infection in adults and children, by **Emory University researchers Dennis Liotta, Raymond Schinazi and Woo-Baeg Choi.** The drug was licensed to Triangle Pharmaceuticals by Emory University in 1996 which was later acquired by Gilead Sciences in 2003.[39] Marketed by Gilead Sciences as Emtriva, this drug received its first FDA approval in July 2003 for the treatment of HIV infection in combination with other antiretroviral agents.

Mechanism of action:

Emtriva works by inhibiting an enzyme called reverse transcriptase that is responsible for copying HIV RNA into new viral DNA (Structure of Emtriva has been provided in Figure 8.7).[40] The moment Emtriva enters into the cells; it is phosphorylated by the cellular enzymes. The phosphorylated form of Emtriva *i.e.* emtricitabine 5′-triphosphate so formed competes with naturally-occurring DNA synthesis substrate deoxycytidine 5′-triphosphate thereby inhibiting the activity of HIV-1 reverse transcriptase. As a result of this process, the number of viral particles or "viral load" in a patient's body is reduced and the number of immune cells (called T-cells or CD4+ T-cells) is increased indirectly.

8.8.1.2.3 Alimta/Pemetrexed (Eli Lilly & Princeton University). A perfect example of how curiosity driven research (an interest in the lovely white, yellow, and red pigments present in the wings of butterflies) has resulted in the synergistic benefit of societal welfare and commercial success is **Alimta – an anticancer drug.** Alimta is a drug approved for the treatment of different types of lung cancer (Figure 8.8) which was discovered by the creative brilliance of Princeton University's Edward C. Taylor in collaboration with a global pharmaceutical company named Eli Lilly & Co. Taylor's fascination with the butterfly wing pigmentation led him to devote much of his research career to the synthesis of antifoliates (compounds that inhibit cell growth by blocking the synthesis of DNA and RNA through targeting essential enzymes present in the wings of butterflies).[41,42] Taylor knew that he could not carry out the task of developing a cancer treatment alone and needed the help of a biochemist. Having served as a consultant to both medicinal chemistry and process development programs at Eli Lilly & Co. in Indianapolis, Indiana, it was quite easy for him to seek this company's approval in getting into the joint venture of developing the drug. As a result of this successful partnering, Alimta became the first drug to be approved by the FDA in 2004 for the treatment of the rare cancer malignant pleural mesothelioma. Within a span of merely five years, it received several other approvals for the cure of advanced non-squamous non-small cell lung cancer as well. This wonderful discovery would never have been possible without an extended commitment towards synthesizing such type of novel compounds with exceptional anticancer properties in Professor Taylor's lab and a true partnership with a company that had the ability to guide development.

Mechanism of action:

Alimta—a pyrrolo-pyrimidine analogue of folic acid—enters into the cells with the help of a reduced folate carrier and high affinity folate receptors, where it gets polyglutamated intracellularly by an enzyme called folylpolyglutamate synthase.[43] As a result of this process, several key enzymes involved in DNA synthesis and folate metabolism are inhibited eventually leading to cell death.

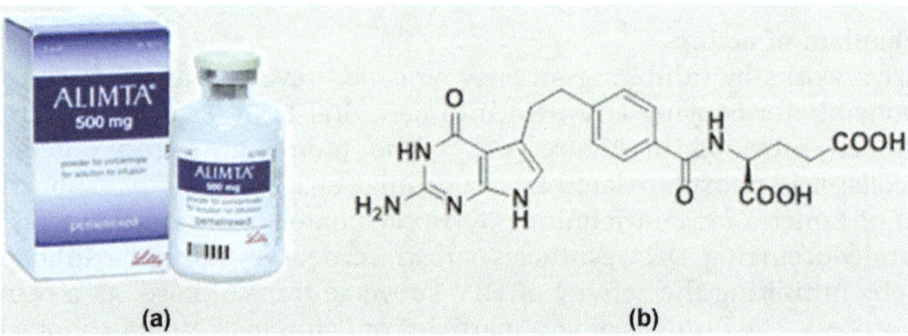

Figure 8.8 (a) Alimta drug manufactured by Eli Lilly and company sold in a powdered form and (b) Structure of the drug.

8.8.1.2.4 Taxol® (Research Triangle Institute and Bristol-Myers Squib). During the latter half of the 20th century, researchers started realizing the need for developing systemic treatments, such as chemotherapy and hormone therapy that would destroy harmful cancerous cells. Consequently, the search for effective anticancer products began far and wide. In 1962, a life-saving natural compound called paclitaxcel (derived from the bark of the Pacific yew tree) was discovered by National Cancer Institute-funded researchers (Figure 8.9).[44]

Systematic investigations of this compound by Dr Monroe E. Wall, Dr Mansukh Wani, and colleagues at the Research Triangle Institute's Natural Product Laboratory revealed that paclitaxel had the potential to cure cancer as the extracts from the bark displayed exceptional cytotoxic activity. However, it took the researchers at the Research Triangle Institute several years to isolate paclitaxel in its pure form. In 1977, Taxol was selected as a candidate for clinical development after the National Cancer Institute was able to confirm antitumor activity in the mouse melanoma B16 model. In 1991 the Research Triangle Institute entered into a partnership with Bristol-Myers Squib for the commercial production of Taxol. Finally, in 1992 the FDA approved Taxol for the treatment of ovarian cancer. Today, Taxol is used for treating different types of cancer such as breast, lung, and ovarian, as well as Kaposi's sarcoma, and is recognized on the WHO's model list of essential medicines.

Mechanism of action:

The mode of action of paclitaxel involves interference with the normal function of microtubules.[45] Specifically, this drug binds to the β-subunit of tubulin which is the building block of microtubules. Due to the binding of palitaxel, the building blocks are locked and the resultant microtubule/palitaxel complex cannot disassemble thus adversely affecting the cell function. Unlike colchicines that result in the depolymerisation of microtubules *in vivo*, palitaxel hyper-stabilizes their structure such that a cell loses its ability to use its cytoskeleton in a flexible manner.

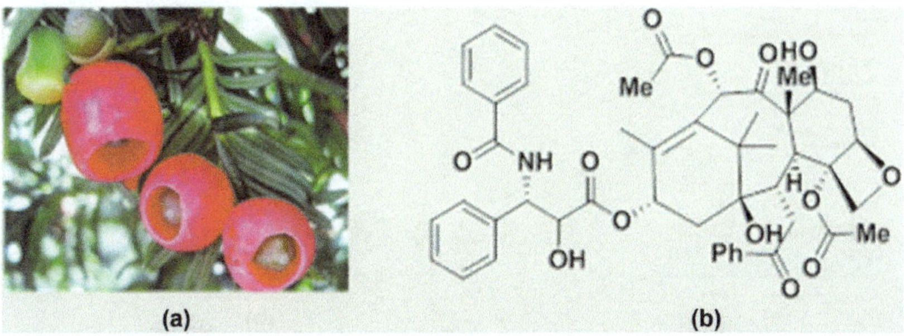

Figure 8.9 (a) Taxus brevifolia: plant that produces Taxol and (b) Structure of paclitaxel also known as Taxol.

8.8.1.2.5 Vorinostat (Columbia University, Sloan Kettering Cancer Center and Merck). Vorinostat, also known as as suberanilohydroxamic acid (SAHA), is the first histone deacetylase (HDAC) inhibitor that was discovered by Columbia University chemist Ronald Breslow and Memorial Sloan Kettering researcher Paul Marks (Figure 8.10).[46] The findings that DMSO causes cancerous red blood cells to lose their disease traits by Charlotte Friend, a microbiologist and oncologist in 1971, ignited a spark of curiosity in Breslow. In a quest to find out how DMSO could cause growth arrest and terminal differentiation of transformed cells, Breslow along with fellow researchers decided to investigate structurally related compounds that might have anticancer effects more potent than those of DMSO. They were able to develop one such potent analogue called SAHA. Although they were not aware of the mechanism of action of this compound, they decided to run with it since it was doing wonderful things with erythroleukemia cells and many other types of cancer cells in culture. Much later in 1991, the first crystal structure of SAHA bound to a histone deacetylase homologue was obtained in collaboration with Sloan Kettering Cancer Center. In 2001, phase I and phase II trials of SAHA and related agents were completed by a small company called Aton Pharma. However, phase III trials could not be achieved hence the compound was decided to be marketed. Thereafter, Merck in 2004 put SAHA into phase III trials. After completing its clinical trials, SAHA was granted FDA approval in October 2006, which it now marketed as vorinostat (Zolinza).

Mechanism of action:

Studies have revealed that vorinostat binds to the active site of histone deacetylases, acting as a chelator for zinc ions that are present in the active site of this enzyme. By forming a strong chelate complex, vorinostat inhibits the activity of histone deacetylases resulting in the accumulation of acetylated histones and acetylated proteins, including transcription factors that are required for the expression of genes to order to induce cell differentiation.[47]

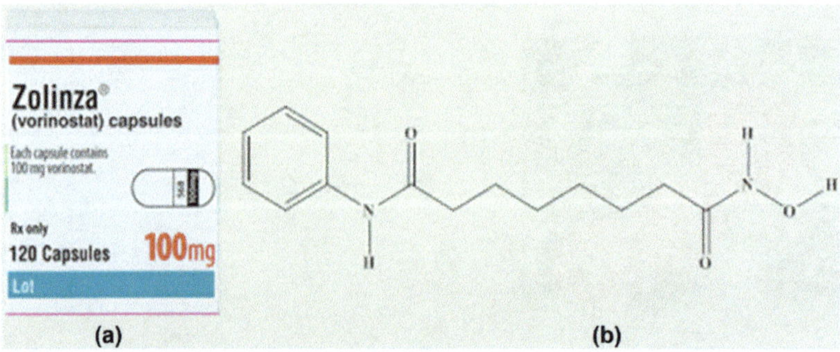

Figure 8.10 (a) Vorinostat (also known as Zolinza) sold in capsule form and (b) Structure of vorinostat.

8.8.2 Design and Development of Novel Pathways for Pharmaceutical Industries

We have discussed a few examples to show how academia has provided a new direction to the industrial researchers, and how a large number of life-saving drugs or new pathways have been discovered with the help of effective academia–industry collaborations.

8.8.2.1 An Efficient Bio-catalytic Process to Manufacture Simvastatin

8.8.2.1.1 Professor Yi Tang Collaborating with CODEXIS. Simvastatin, a synthetic analogue of lovastatin (a fungal natural product), is a leading therapeutic drug used for the treatment of hypercholesterolemia and diabetic cardio-myopathy.[48] Originally developed and marketed by Merck under the brand name "Zocor", simvastatin became the largest selling drug for lowering cholesterol worldwide with its sale accounting for about $5 billion. Traditional synthetic routes to simvastatin were based on the conversion of lovastatin to simvastatin through the addition of a methyl group which required protection and subsequent deprotection of functionalities present in the lovastatin molecule. Unfortunately, these multistep synthetic methodologies were not only uneconomical, but also wasteful as they involved enormous amounts of toxic and hazardous reagents. Besides, despite considerable optimization, the product yield of the desired drug molecule could not be increased beyond 70%. However, Professor Yi Tang and his group from University of California, Los Angeles (UCLA) came up with an excellent bio-catalytic strategy that circumvented tedious protection, deprotection steps resulting in greater atom economy, and minimum waste generation.[49] They used a process called "directed evolution" to create an engineered enzyme (cloned LovD, a natural acyltransferase produced by *Aspergillus terreus* involved in synthesizing lovastatin) that could be employed for carrying out the initial synthesis of simvastin. **"Codexis"**—a leading developer of biocatalysts for pharmaceutical and fine chemical industries—obtained the license of this intellectual technology from UCLA and eventually optimized the enzyme as well as the chemical process for commercial manufacture of this drug.[50]

Previously, simvastatin was synthesized *via* the hydrolysis of lovastatin molecule to monacolin J which was followed by protection, esterification and deprotection steps. The new route on the other hand involved the quantitative hydrolysis and conversion of lovastatin into the water-soluble ammonium salt of monacolin J followed by its regioselective acylation at the C8 position using a genetically evolved variant of LovD acyltransferase (Figure 8.11). Methyl 3-mercaptopropionic acid is the only co-product obtained in this process which can be recycled back. This novel strategy not only reduces the number of steps required for the manufacture of simvastatin from four to two but also increases its yield from a mere 70% to 97% at

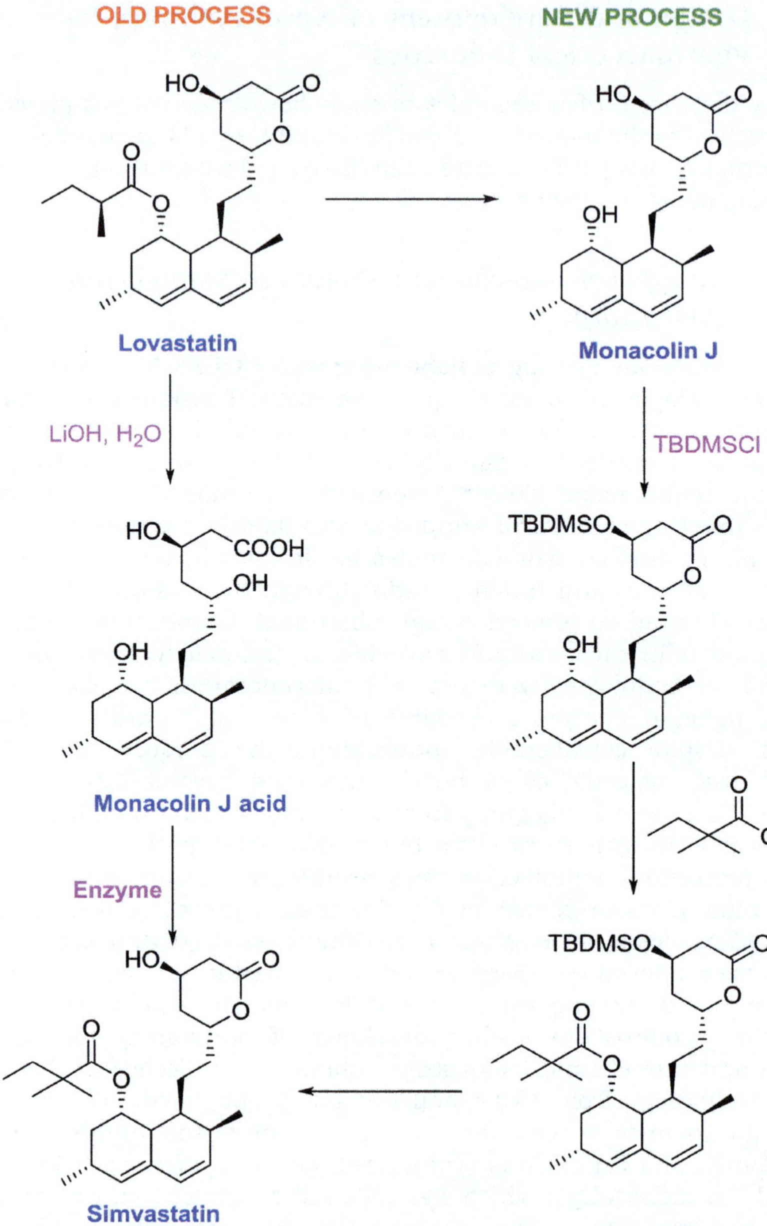

Figure 8.11 A schematic comparison of old and new methods of Simvastin synthesis.

a loading of only 5 grams per liter of monacolin J. In addition, it avoids the use of toxic reagents like *tert*-butyl dimethyl silane chloride, methyl iodide, and *n*-butyl lithium that are employed in the traditional method of simvastatin synthesis. The efficiency of this process can be judged from the fact that every year 10 metric tons of silvastatin are synthesized using this technique.

8.8.2.2 Modified Green Route for the Manufacture of Ranitidine

Ranitidine HCl is world's highest selling anti-histamine drug that is used for the treatment and prevention of peptic ulcers, gastro-esophageal reflux and Zollinger–Ellison syndrome.[51,52] Every year, nearly 4000 metric tonnes of ranitidine HCl (greater than 50% of the world's requirement) is manufactured by Indian Pharmaceutical industries. During the commercial production of this drug, an obnoxious-smelling gas called "*methyl mercaptan*" is emitted as waste (Figure 8.12).

Methyl mercaptan is a highly toxic, colourless low-boiling liquid that undergoes rapid conversion into the gaseous state at ambient temperature and pressure conditions.[53] The low vapour density of this gas enables its facile diffusion into the atmospheric air which contaminates it and creates air that is foul-smelling. In fact, genotoxic studies reveal that prolonged exposure to a high concentration of this gas may lead to several fatal disorders like central nervous system damage, malfunctioning of respiratory organs and even death in certain cases. Considering the gravity of the situation, many of the manufacturing plants in India were asked to shut down their production by the pollution control boards. Incineration has been widely adopted by some renowned pharma industries manufacturing this life-saving drug, such as Nuland Laboratories, Dr Reddy's Lab., Saraca Laboratories and SMS Pharmaceuticals Ltd., as a solution to handle problems related with the release of this harmful gas. However, this technique suffers from a number of drawbacks. Firstly, it leads to the generation of flue gases such as sulfur dioxide (SO_x) which causes environmental pollution. Secondly, another frequently encountered problem of this technique is the need to regulate the combustion rate to control the volume of emissions which is generally not uniform as drugs are synthesized using a batch method. Therefore, industries have been looking for an eco-friendly as well

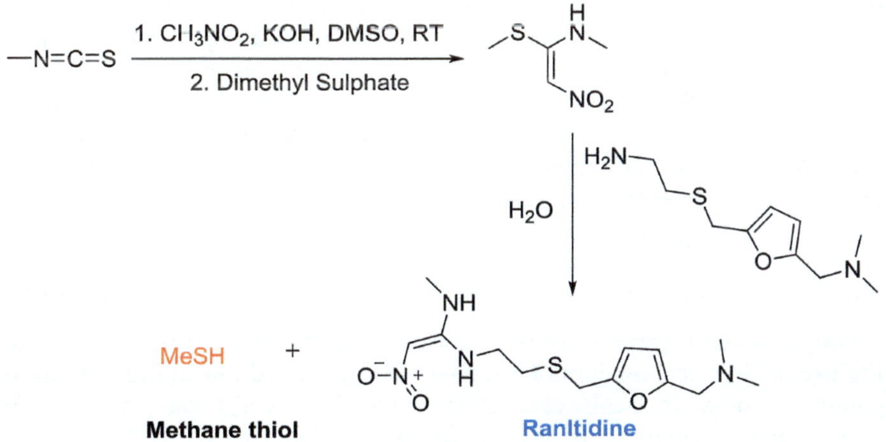

Figure 8.12 Commercial production of Ranitidine HCl.

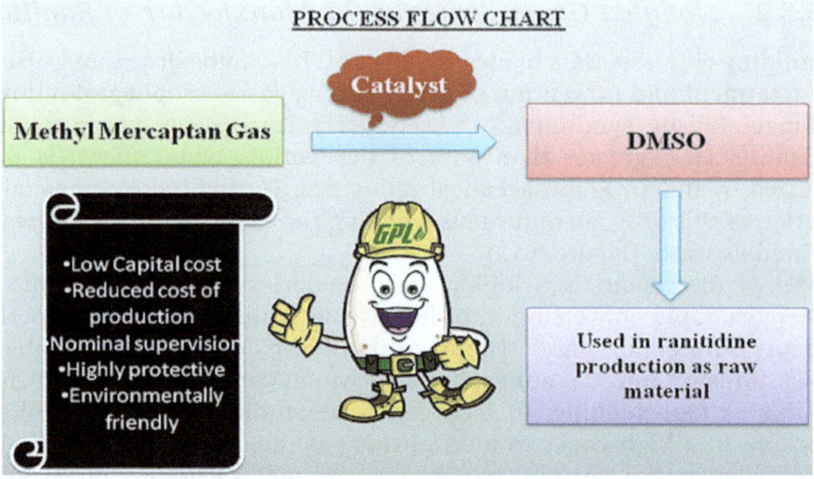

Figure 8.13 Process flow chart depicting the green catalytic conversion of methyl mercaptan gas.

as cost effective solution that would meet the challenges of ranitidine synthesis.

To combat issues concerning ranitidine synthesis, Professor Mihir K. Chaudhuri of IIT Guwahati and SMS pharmaceuticals Ltd., Hyderabad came up with an innovative solution.[39] They developed an environmentally benign catalytic protocol for converting the foul-smelling methyl mercaptan gas into the value added DMSO (one of the starting materials utilized for the ranitidine manufacturing process shown in Figures 8.13 and 8.14).

The DMSO so generated can be recycled back into the process which reduces the cost of ranitidine production by as much as 40%. This novel protocol has been utilized by industries for the commercial production of the drug using mostly vanadium–titanium or titanium–phosphorous based compounds as catalysts. It is worth mentioning here that the synergistic integration of the rich academic experience of Professor Chaudhuri of IIT Guwahati and the industrial expertise of Mr P. Ramesh Babu and Mr T. V. Srihari of SMS Pharma Ltd. and Mr Suresh Babu of RCHEM Ltd. lead to the commercial success.

8.8.2.3 Aerobic Oxidation Methods for Pharmaceutical Synthesis

Amongst several organic transformations, oxidation reactions that involve the use of chemical oxidants have been widely carried out in industry for the generation of synthetically valuable products. However, these reactions often employ toxic oxidants, involve harsh reaction conditions and lead to the formation of wasteful by-products. To overcome the drawbacks of such

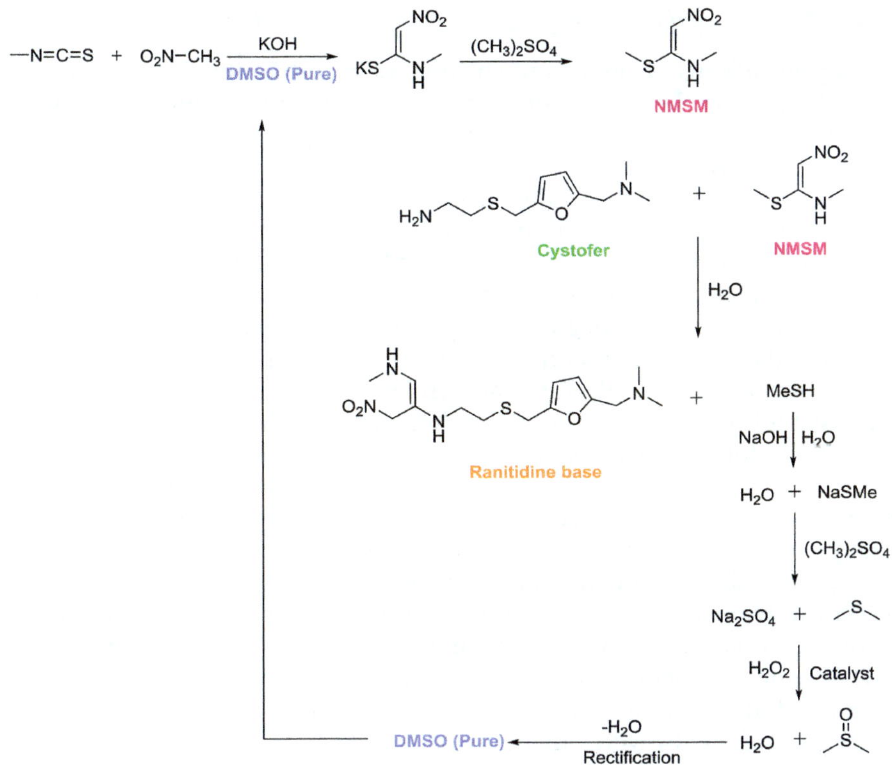

Figure 8.14 Reaction scheme illustrating the conversion of methyl mercaptan gas into DMSO.

oxidative processes, Professor Stahl designed a catalytic methodology that replaced the use of hazardous oxidizing agents with an environmentally benign oxidant – "molecular oxygen from air."[54–56] Experimental results revealed that this new oxidation process not only worked under mild conditions, but also could be performed safely on a large scale saving time and money and also reducing hazardous waste. The development of this new practical and safe method has opened the gateway for benign large-scale industrial oxidative synthesis.

In fact, Professor Stahl and his group members have been continuously working in this direction and for the scalable implementation of these reactions for pharmaceutical synthesis they have entered into partnership with Professor Thatcher Root who works at the Department of Chemical and Biological Engineering, University of Wisconsin-Madison, and scientists at several pharmaceutical companies such as Martin Johnson, Eli Lilly, Merck and Pfizer. Working with the industrialists, a continuous flow process has been developed which has been used to achieve aerobic oxidation of alcohols to aldehydes in near-quantitative yields with reactor residence times as low as five minutes.

8.8.2.4 Towards Ending Our Dependence on Organic Solvents

Every year billions of pounds of volatile, toxic and flammable organic solvents are used by chemical manufacturers to carry out organic reactions that eventually end up as waste. Water could be thought of as a promising green alternative, however unfortunately it cannot replace organic solvents as many chemical reagents do not either dissolve or react in water. Professor Lipshutz from University of California, Santa Barbar has designed a second-generation surfactant called **TPGS-750-M** which is primarily composed of inexpensive ingredients like tocopherol, succinic acid and methoxy poly(ethylene glycol) – a degradable hydrophilic group also known as MPEG-750 for enhancing the solubility of organic reactants in water.[57,58] The surfactant forms nanomicelles in water (micelles are lipophilic on the inside and hydrophilic on the outside) that then act as nanoreactors by facilitating the dissolution of reactants and catalysts, due to which an overall increase in reaction rates is observed. It has been found that a number of organic reactions, including ruthenium-catalysed olefin metatheses, palladium-catalysed cross-couplings such as Suzuki, Heck and Sonogashira, aminations, silylations and borylations, have been successfully carried out within TPGS-750-M micelles in water at room temperature and the desired products have been obtained with relatively high yield (Figure 8.15). Recently, using this designed surfactant, new C–C bonds have been constructed *via* aromatic C–H bond activation at room temperature which is indeed a great achievement. What is really worth mentioning here is that:

- No frothing or foaming is associated with this surfactant which makes product isolation quite straightforward

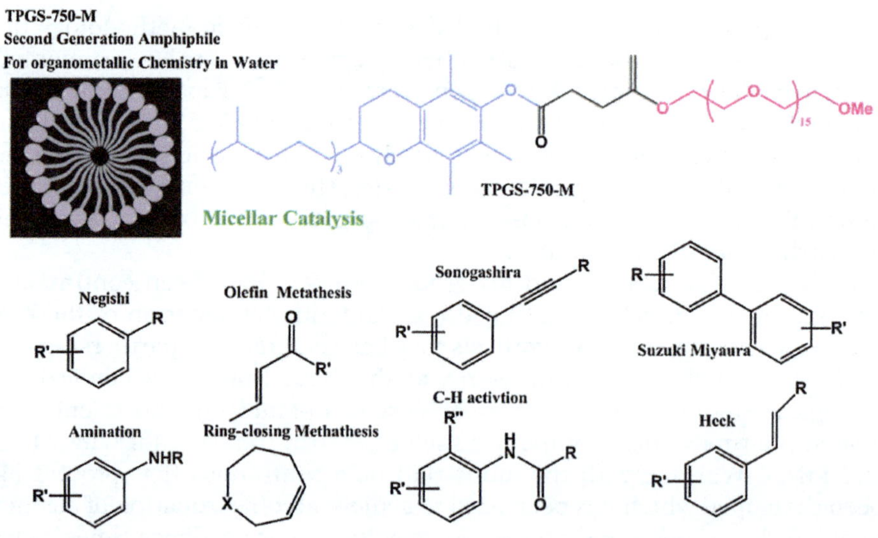

Figure 8.15 Micellar catalysis using the second-generation surfactant designed by Prof. Lipshutz.

- The surfactant can be recycled for several cycles after use without any appreciable loss in its activity and the insoluble product can be recovered simply by means of extraction
- Reactions can be done in seawater as high-quality water is not required in this case

Thus, Professor Lipshutz's technology offers a promising pathway to the industrial chemists to perform organic reactions without using toxic organic solvents. Currently, Sigma–Aldrich, an American chemical, life science and biotechnology company owned by Merck KGaA, is offering this surfactant "TPGS-750-M" for sale, therefore making it widely available to research laboratories.

8.8.2.5 Recycling CO_2 to Synthesize Higher Alcohols

Higher alcohols containing three to eight carbon atoms are of interest since they can be used as chemical feed-stocks and transportation fuels.[59] As fuel substitutes they possess several advantages over small chain alcohols, including higher energy density, lower hygroscopicity and lower vapour pressure that ultimately leads to better air quality. Bio-synthesis of higher chain alcohols from CO_2 represents the most promising (because of its capability to reduce net carbon emissions) yet very challenging route (as naturally-occurring microorganisms cannot produce these alcohols). It was for the very first time that Dr James Liao, a professor at the UCLA and a pioneer in Metabolic Engineering, Synthetic Biology, and Systems Biology designed a genetically engineered microorganism for synthesizing higher alcohols from glucose or directly from CO_2 and even won a Presidential Green Chemistry Challenge Award in 2010 for his incredible work (Figure 8.16).[60] Using this

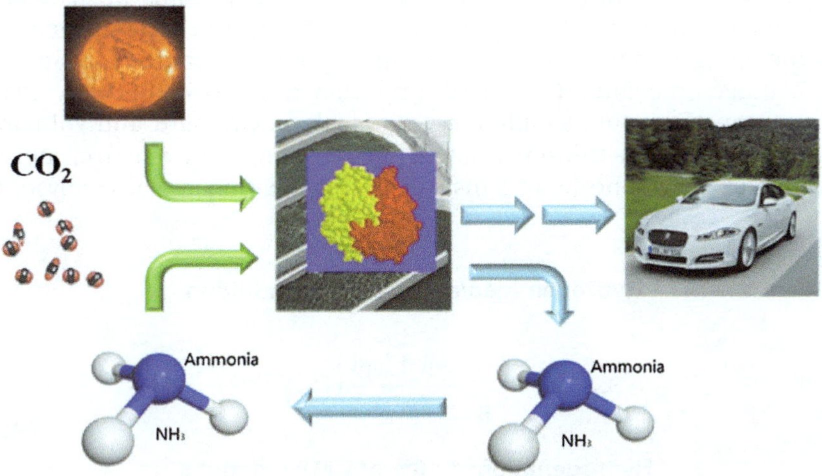

Figure 8.16 Next-generation biofuels, CO_2 fixation, electricity to fuels, cellulosic fuels, protein bio-refining.

microbial technology, Professor Liao and his team members have been able to generate isobutanol from glucose in quantitative yields with high efficiency and specificity. Dr Liao's technology has proven to be highly promising in converting solar energy and CO_2 into chemical feed-stocks and has been able to eliminate 500 million tons of CO_2 emissions. Now, Easel Biotechnologies, co-founded by Dr James Liao and NantWorks LLC in 2010, is commercializing this CO_2-to-fuels technology under exclusive license from UCLA.

8.8.2.6 Hydrogen-mediated Carbon–Carbon Bond Formation for the Synthesis of Important Pharmaceuticals with Minimal Waste Generation

Ever since the first construction of a carbon–carbon (C–C) bond by Kolbe in the year 1845, the C–C bond formation reactions have played a key role in expanding our horizon of chemical synthesis by assembling complex molecular frameworks.[61,62] Although these reactions lead to the generation of industrially significant complex molecules with utmost ease, one cannot ignore the enormous amount of waste generated during the manufacturing of such compounds. Considering the need for developing waste free methodology for creating new C–C bonds, Professor Michael J. Krische and his group at the University of Texas at Austin designed and synthesized a new class of metal-catalysed hydrogenation reactions that enable the construction of C–C bonds in a highly selective fashion with utmost ease and minimal waste generation.[63,64] Unlike the classical hydrogenation reactions wherein reductions mediated by hydrogen lead to the formation of C–H bonds, Krische's approach involves the combination of two or more organic molecules with hydrogen gas in presence of a suitable metal catalyst to generate a C–C bond (Figure 8.17). Because all the atoms present in the reactants appear in the final product, no by-product or waste is generated which eliminates pollution at source. Apart from this, another advantage of this hydrogen-mediated C–C bond formation reaction is that it avoids the use preformed organometallic reagents (such as Grignard and Gilman reagents in carbonyl addition reactions) which are highly reactive, moisture sensitive and pyrophoric and instead uses transition metal catalysts that

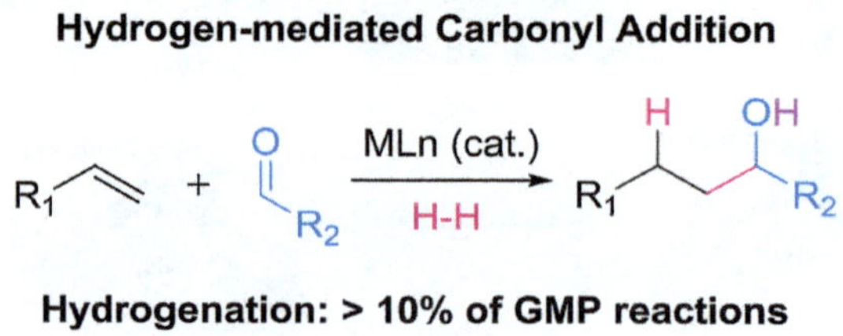

Figure 8.17 Prof. Krische's approach for C–C bond formation.

avoid the hazards of traditional organometallic reagents. This example represents the first attempt of exploiting hydrogenation in C–C couplings reactions beyond the conventional alkene hydroformylation and Fischer–Tropsch reaction. Certainly, Krische's research has added a new dimension to one of chemistry's most fundamental catalytic processes as it has the potential to eliminate vast quantities of hazardous chemicals.

8.9 Redefining the Operational Methods of Global Industries Towards Sustainability

8.9.1 Green Solvents

8.9.1.1 Industrial Research

While greening industries, solvents have always been a key concern since they are used in large volumes, are typically volatile organic compounds, and they have a high risk for waste generation, air pollution, and other health issues. Therefore, it is imperative to find either safer and more efficient alternatives or remove solvents altogether. To date, many Presidential Green Chemistry Awards have recognized approaches that reduce the use of solvents; these include alternative processes that use either green solvents like carbon dioxide or water or avoid the use of solvents completely.[65,66]

Conventionally, water is not considered as a suitable medium for organic reactions due to limited solubility of many substrates and its reactivity with a variety of functional groups. But lately, due to clean chemistry there has been a revival of interest in chemistry under aqueous medium. Alternatively, supercritical carbon dioxide is an inexpensive and environmentally benign solvent. A variety of supercritical fluids (SCFs) have been used for diverse applications such as extraction, synthesis, chromatography, catalysis, dry cleaning, *etc.*[67] The use of such solvents works under the concept of simple pressure release leading to solvent-free clean products and non-toxic CO_2 that can be easily recycled. A recent example of a green solvent is supercritical carbon dioxide, which functions like a liquid under pressure and at slightly elevated temperature. This has been commercially employed in fabric dyeing by a process invented by DyeCoo Textile Systems at industrial scale.[68] This has replaced the traditional dyeing that requires a lot of water to dye a single T-shirt (about 7 gallons) and is energy intensive. For both environmental and human health reasons, manufacturers of spray cleaners, laundry detergents, and other cleaning products for home and industry have been replacing solvents with greener solvents, taking into account their high performance. Recently, both Procter & Gamble and DuPont announced plans to use cellulosic ethanol derived from corncobs and stalks in "TideColdwater Clean" which will replace ethanol derived from corn kernels. This will not only help to reuse agricultural waste but will save the energy required to wash clothes. Another bio-based solvent sold by DuPont is 1,3-propanediol, which acts as stabilizer, and enzyme carrier. It is found in spray cleaners and laundry detergents. Ethyl levulinate glycerol ketal is another green solvent manufactured by the bio-based chemicals start-up Segetis which helps solubilize fragrance oils and keep the overall cleaning formula stable. Butyl 3-hydroxybutyrate also works as a green cleaning solvent, developed by Eastman Chemical. In addition, Elevance Renewable Sciences, Inc. converts vegetable oils into specialty chemicals with olefin metathesis technology invented by Nobel laureate Robert H. Grubbs, to produce two green solvents.[73] Along with surfactants manufacturer Stepan, this company produced a surfactant that replaces solvents like *n*-methylpyrrolidone and methylene chloride in adhesive removers and paint strippers.[69]

8.9.1.2 *Academic Research*

A team at the University of Wisconsin recently developed a promising bio-based green solvent using mineral acid catalysts for the conversion of hemicellulose and cellulose biomass into high value platform chemicals and transportation fuels. The use of such solvents minimized side-reactions in the aqueous phase and also enabled recycling of the mineral acid catalysts.[70]

8.9.2 Next-generation Catalyst Design

8.9.2.1 *Industrial Research*

Recently, twenty Presidential Green Chemistry Challenge Awards have recognized green catalysts. Example include technologies developed by Elevance that uses a Nobel-prize-winning catalysis approach to produce high-performing, green specialty chemicals at advantageous costs. This technology breaks down natural oils and then recombines those fragments into high-performance green chemicals which blends the benefits of both bio-based products and petrochemicals. In comparison to petrochemical technologies, this technology consumes considerably lesser energy and reduces greenhouse gas emissions by 50%.[71] Another green catalyst recognized by Dow Chemical's awards was found to reduce the environmental footprint associated with propylene oxide production. This process was developed with the collaboration of BASF, to produce moieties needed to generate vast array of products including food additives, detergents, polyurethanes, and personal care items. Reduction in wastewater production and energy consumption was observed to a great extent over traditional technologies.[72] Lately, another catalyst containing manganese-mullite materials with samarium or gadolinium was developed for cleaning diesel engine exhaust. It proved to be less expensive and more efficient in converting toxic diesel engine exhaust product nitric oxide to a more benign nitrous oxide.[73]

8.9.2.2 Academic Research

A powerful series of tetra-amido macrocyclic ligand (TAML) catalysts modelled on natural peroxidase enzymes was developed at Carnegie Mellon University. This green catalyst was found to have potential in reducing the impact of pharmaceutical industries on the environment by breaking down a wide variety of chemical residues (including those from Lipitor, Prozac, Zoloft, the contraceptive pill) in the sewage treatment process.[74]

8.9.3 Microwave-assisted Synthesis

8.9.3.1 Industrial Research

During the past few years, there has been an upsurge interest in the use of microwave irradiation within the pharmaceutical industry to promote synthetic transformations that aid in the production of new chemical entities/drugs.[75,76] The factors that have worked as a catalyst towards accelerating the uptake of this technology include broad applicability: both solution as well as solid phase reactions; increase in reaction rates up to 1000-fold in the best cases due to reduction in time required for carrying out the synthesis as compared to the traditional conductive heating methods; dramatic increase in the product yields; scalable from sub-milligram to multigram quantities; green chemistry reactions in supercritical water or solvent-less reactions; controlled method of heating; rapid reaction optimization, *etc.* Principally, pharmaceutical companies have utilized this technology in three areas of drug research: (i) the screening of organic drug formulae; (ii) peptide synthesis; and (iii) DNA amplification. Several life-saving drugs have been synthesized using microwave reactors. Examples include the synthesis of aspirin (using a catalyst solvent-free approach under microwave irradiation),[77] Imatinib, a blockbuster drug used to treat certain types of cancer (using speedy, high yield and convenient synthesis using an aldehydic, super acid-sensitive resin, through an efficient, microwave-assisted synthetic protocol) and many more.[78]

8.9.3.2 Academic Research

A powerful microwave-assisted synthetic approach that utilizes neat reactants either in the presence of a catalyst or catalysed by the surfaces of recyclable support materials such as alumina, silica, clay, and 'doped' surfaces such as $NaIO_4$–silica, iron(III) nitrate–clay (clayfen), and persulfate–clay was reported by Dr Rajender S. Varma from Sam Houston State University.[79] The technology addresses the needs of a broad chemical community, such as polymers, pharmaceuticals, and fine chemicals *via* efficient production of valuable intermediates (enones, imines, enamines, nitroalkenes, oxidized sulfur species and heterocycles). Dr Varma has been recognized by EPA for his great contribution to the field of organic synthesis.

8.9.4 Flow Chemistry

8.9.4.1 Industrial Research

In recent times, the introduction of flow chemistry has completely transformed the landscape of synthetic organic chemistry by changing the way in which reactions are performed at industrial and research levels. Some of the outstanding features associated with flow reactors include improved thermal management, enhanced mass transfer, mixing control, scalability, rapid chemical reactions and ability to withstand harsh reaction conditions. The multiple benefits of flow chemistry have also been realized tremendously by many of the researchers from big pharmaceutical companies like Bristol-Myers Squibb, Novartis Pharma *etc.* In particular, scientists at Novartis Pharma designed a combination of batch and flow technologies and successfully synthesized the anti-diabetic drug "vildagliptin."[80] For the synthesis of this drug usually a Vilsmeier reaction is utilized, wherein substituted amides are reacted with phosphorous oxychloride and electron-rich aromatic compounds and aryl aldehydes and ketones are. The large-scale applicability of this reaction is hindered as it is associated with a serious drawback – the formation of a highly hazardous Vilsmeier reagent. However, when flow chemistry was employed, the in-line formation and immediate consumption of Vilsmeier reagents was achieved and thus the problem could be resolved.

8.9.4.2 Academic Research

In 2013, Steven Ley and co-workers came up with a new flow platform technology which was based on fluoropolymer peristaltic pumps (those ideal for operating with highly reactive and air-sensitive compounds like

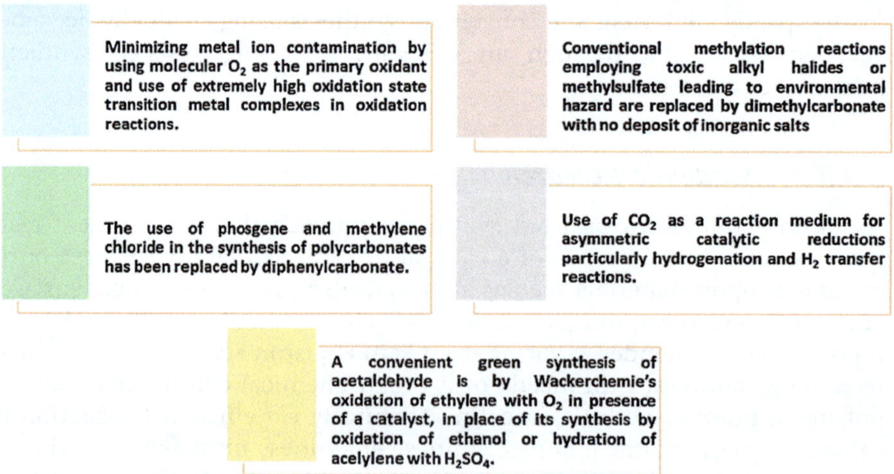

Figure 8.18 Examples of hazardous reagent substitution.

organolithium and Grignard reagents) for the multistep synthesis of tamoxifen.[81] Tamoxifen is an antagonist prodrug that is utilized for treating breast cancer. This use of this new flow procedure showed very high efficiency as it combined four chemical reactions into one stream, thereby minimizing manual intervention and reducing risks associated with the handle of organometallic reagent.

As already mentioned previously, research is ongoing in both industrial as well as academic setting to develop alternative green approaches that enable important organic transformations to be performed safely on a commercial scale. Some of these approaches include development of safer forms of hazardous reagents, identification of new synthetic conditions, such as flow and microwave chemistry *etc.* Figure 8.18 highlights examples of hazardous reagent substitution practices that have been followed for improving the green credentials of different reactions.

8.10 Conclusion

There is little doubt that progress in modern scientific research can be achieved only through effective teamwork of academia and industry. Success stories of several blockbuster drugs discovered with the help of academic intelligence and industrial support highlight this fact. Owing to the benefits associated with such partnerships, leading pharmaceutical companies like GlaxoSmithKline, AstraZeneca, Pfizer *etc.* have entered into close collaborations with different universities as they are well aware of the fact that academic researchers can show them the right path to tackle the challenges of drug discovery and development through their knowledge and expertise. Despite of the fact that industry–academia collaborations have been markedly rewarding and have increasingly benefitted both parties, many are reluctant to engage themselves into such close relationships. Of course, they have their own reasons: either it is the difference in the thought processes of the industrialists and academicians or it is the policy concerning collaborations or irreproducibility of data and so forth. So, there is a clear need to firstly identify the barriers and then come up with effective solutions. Looking to the future, the next important trend is to spread the sustainability mindset across the industry and academia, so as to prepare the next generation of green chemists. The key to achieving the goal of reducing the generation of toxic waste and the use of hazardous solvents and reagents is their widespread substitution by greener alternatives. This can be done by utilizing the simplest raw materials such as H_2, O_2, H_2O, H_2O_2, NH_3, CO and CO_2 in chemical processes. Another powerful tool in chemical industry is catalysis which has led to the evolution of synthesis processes. Likewise, the widespread substitution of classical mineral and Lewis acids by heterogeneous acid catalysts and the introduction of recyclable solid bases will result in a dramatic reduction of inorganic waste. Besides this, phasing out of deleterious organic solvents and their benign replacements will lead to significant contribution towards industrial green chemistry. Also, the use of microwave and flow reactors will play an instrumental role in greening industries. In this chapter, we have discussed the benefits of green chemistry, need for academia–industry collaborations during the present times, the impediments to collaborations, fruitful keys that can be adopted by both the sectors. We believe that through a better understanding on greener perspectives and mutual respect for each other, pharmaceutical companies and academic institutions can engage in stable collaborations that can lead to more scientific discoveries, faster drug development and better health care treatments.

References

1. J. Kling, Academia and the company coin, *Nat. Biotechnol.*, 2009, **27**(5), 411–414.
2. I. Khanna, Drug discovery in pharmaceutical industry: productivity challenges and trends, *Drug Discovery Today*, 2012, **17**(19), 1088–1102.
3. M. Rosenblatt, How Academia and the Pharmaceutical Industry Can Work Together: The President's Lecture, Annual Meeting of the American Thoracic Society, San Francisco, California, *Ann. Am. Thorac. Soc.*, 2013, **10**(1), 31–38.
4. N. S. Gray, Drug discovery through industry-academic partnerships, *Nat. Chem. Biol.*, 2006, **2**(12), 649–653.
5. C. Lengauer, L. A. Diaz and S. Saha, Cancer drug discovery through collaboration, *Nat. Rev. Drug Discovery*, 2005, **4**(5), 375–380.
6. J. H. Glyn, Personal views: The discovery of cortisone: a personal memory, *Br. Med. J.*, 1998, **317**(7161), 822.
7. H. B. Woodruff, Selman A. Waksman, winner of the 1952 nobel prize for physiology or medicine, *Appl. Environ. Microbiol.*, 2014, **80**(1), 2–8.
8. R. M. Rydzewski, *Real World Drug Discovery: A Chemist's Guide to Biotech and Pharmaceutical Research*, Elsevier, 2010.
9. J. Chin-Dusting, J. Mizrahi, G. Jennings and D. Fitzgerald, Finding improved medicines: the role of academic–industrial collaboration, *Nat. Rev. Drug Discovery*, 2005, **4**(11), 891–897.
10. C. A. Challener, New ways around hazardous reagent chemistry, *Pharm. Technol.*, 2015, **39**(6), 34–36.
11. https://www.navigantresearch.com/newsroom/green-chemical-industry-to-soar-to-98-5-billion-by-2020.
12. D. C. Mowery, R. R. Nelson, B. N. Sampat and A. A. Ziedonis, The growth of patenting and licensing by US universities: an assessment of the effects of the Bayh–Dole act of 1980, *Res. Policy*, 2001, **30**(1), 99–119.
13. C. J. Tralau-Stewart, C. A. Wyatt, D. E. Kleyn and A. Ayad, Drug discovery: new models for industry–academic partnerships, *Drug Discovery Today*, 2009, **14**(1), 95–101.
14. P. D'este and M. Perkmann, Why do academics engage with industry? The entrepreneurial university and individual motivations, *J. Technol. Transfer*, 2011, **36**(3), 316–339.

15. J. Bruneel, P. d'Este and A. Salter, Investigating the factors that diminish the barriers to university–industry collaboration, *Res. Policy*, 2010, **39**(7), 858–868.
16. B. H. Hall, A. N. Link and J. T. Scott, Barriers inhibiting industry from partnering with universities: evidence from the advanced technology program, *J. Technol. Transfer*, 2001, **26**(1–2), 87–98.
17. M. Perkmann and K. Walsh, University–industry relationships and open innovation: Towards a research agenda, *Int. J. Manage. Rev.*, 2007, **9**(4), 259–280.
18. E. M. Greitzer, J. A. Pertuze, E. S. Calder and W. A. Lucas, Best practices for industry-university collaboration, *MIT Sloan Manage. Rev.*, 2010, **51**(4), 83.
19. T. Sparey and F. Gliubich, Learning from each other, *Nat. Biotechnol.*, 2008, **26**(5), 491–494.
20. A. Shapira and Y. Rosenfeld, Achieving construction innovation through academia-industry cooperation—Keys to success, *J. Prof. Issues Eng. Educ. Pract.*, 2010, **137**(4), 223–231.
21. UN ESCAP, K. O. I. C. A. (2012). Low Carbon Green Growth Roadmap for Asia and the Pacific. *Urban Transport: policy recommendations for the development of eco-efficient infrastructures, Bangkok, Thailand.*
22. J. L. Tucker and M. M. Faul, Drug companies must adopt green chemistry, *Nat. News*, 2016, **534**, 27–28.
23. J. F. Jenck, F. Agterberg and M. J. Droescher, Products and processes for a sustainable chemical industry: a review of achievements and prospects, *Green Chem.*, 2004, **6**(11), 544–556.
24. C. Santi, C. Scalera, C. Tidei, V. Dragone, L. Incipini and L. Di Schino, 2011, Green Chemistry for Environmental Sustainability: An Example of "Bio-Logic" Approach. In *1st World Sustainability Forum (Proceedings in Sustainability).*
25. S. Everts, Better living through green chemistry: Pharmaceuticals, *New Sci.*, 2010, **15**, 57.
26. P. J. Dunn, S. Galvin and K. Hettenbach, The development of an environmentally benign synthesis of sildenafil citrate (Viagra™) and its assessment by Green Chemistry metrics, *Green Chem.*, 2004, **6**, 43–48, DOI: 10.1039/B312329D.
27. U.S. Environmental Protection Agency. Presidential Green Chemistry Challenge Awards: 2006 Greener Reaction Conditions Award. http://www2.epa.gov/green-chemistry/2006-greener-reaction-conditions-award (accessed June 30, 2015).
28. U.S. Environmental Protection Agency. Presidential Green Chemistry Challenge Awards: 1997 Greener Synthetic Pathways Award. http://www2.epa.gov/green-chemistry/1997-greener-synthetic-pathways-award (accessed June 30, 2015).
29. U.S. Environmental Protection Agency. Presidential Green Chemistry Challenge Awards: 2012 Greener Synthetic Pathways Award. http://www2.epa.gov/green-chemistry/2012-greener-synthetic-pathways-award (accessed June 30, 2015).

30. American Chemical Society. Examples of Green Chemistry. http://www.acs.org/content/acs/en/greenchemistry/what-is-green-chemistry/examples.html (accessed June 30, 2015).
31. U.S. Environmental Protection Agency. Presidential Green Chemistry Challenge Awards: 2004 Greener Synthetic Pathways Award. http://www.epa.gov/greenchemistry/pubs/pgcc/winners/gspa04.html (accessed June 30, 2015).
32. U.S. Environmental Protection Agency. Presidential Green Chemistry Challenge Awards: 2010 Greener Reaction Conditions Award. http://www2.epa.gov/green-chemistry/2010- greener-reaction-conditions-award (accessed June 30, 2015).
33. S. K. Ritter, EPA Analysis Suggests Green Success, *Chem. Eng. News*, 2015, **93**(5), 32–33.
34. J. Frearson and P. Wyatt, Drug discovery in academia: the third way? *Expert Opin Drug Discovery*, 2010, **5**(10), 909–919.
35. A. K. Ghosh, Z. L. Dawson and H. Mitsuya, Darunavir, a conceptually new HIV-1 protease inhibitor for the treatment of drug-resistant HIV, *Bioorg. Med. Chem.*, 2007, **15**(24), 7576–7580.
36. K. H. Busse and S. R. Penzak, Darunavir: A second-generation protease inhibitor, *Am. J. Health-Syst. Pharm.*, 2007, **64**, 15.
37. D. C. Liotta and G. R. Painter, Discovery and Development of the Anti-Human Immunodeficiency Virus Drug, Emtricitabine (Emtriva, FTC), *Acc. Chem. Res.*, 2016, **49**(10), 2091–2098.
38. P. Ammaranond and S. Sanguansittianan, Mechanism of HIV anti-retroviral drugs progress toward drug resistance, *Fundam. Clin. Pharm.*, 2012, **26**(1), 146–161.
39. E. C. Taylor, From the Wings of Butterflies, *Chem. Int.*, 2011, **33**(5), 4–9.
40. E. C. Taylor, The Discovery of Alimta (Pemetrexed), *Success. Drug Discovery*, 2015, 157–180.
41. A. A. Adjei, Pemetrexed (ALIMTA), a novel multitargeted antineoplastic agent, *Clin. Cancer Res.*, 2004, **10**(12), 4276s–4280s.
42. R. C. Donehower, The clinical development of paclitaxel: a successful collaboration of academia, industry and the National Cancer Institute, *Oncologist*, 1996, **1**(4), 240–243.
43. S. B. Horwitz, Taxol (paclitaxel): mechanisms of action, *Ann. Oncol.*, 1993, **5**, S3–S6.
44. S. Grant, C. Easley and P. Kirkpatrick, Vorinostat, *Nat. Rev. Drug Discovery*, 2007, **6**(1), 21–22.
45. R. Sarfstein, I. Bruchim, A. Fishman and H. Werner, The mechanism of action of the histone deacetylase inhibitor vorinostat involves interaction with the insulin-like growth factor signaling pathway, *PLoS One*, 2011, **6**(9), e24468.
46. P. A. Todd and K. L. Goa, Simvastatin, *Drugs*, 1990, **40**(4), 583–607.
47. X. Xie, W. W. Wong and Y. Tang, Improving simvastatin bioconversion in Escherichia coli by deletion of bioH, *Metab. Eng.*, 2007, **9**(4), 379–386.

48. K. K. Rana and S. Rana, Fundamentals, Representative Applications and Future Perspectives of Green Chemistry: A Short Review, *Open Access Library J.*, 2014, **1**(05), 1.
49. E. R. Berndt, L. Bui, D. R. Reiley and G. L. Urban, Information, marketing, and pricing in the US antiulcer drug market, *Am. Econ. Rev.*, 1995, **85**(2), 100–105.
50. R. Wright, How Zantac became the best-selling drug in history, *Mark. Health Serv.*, 1996, **16**(4), 24.
51. R. G. Glushkov, E. V. Adamskaya, T. I. Vosyakova and A. F. Oleinik, Pathways of synthesis of ranitidine (review), *Pharm. Chem. J.*, 1990, **24**(5), 369–373.
52. P. M. Patel, A. A. Patil, M. D. Patil, P. S. Patil and S. L. Borse, Green Chemistry-An Overview, *Asian J. Res. Chem.*, 2013, **6**, 7.
53. S. S. Stahl, Palladium oxidase catalysis: selective oxidation of organic chemicals by direct dioxygen-coupled turnover, *Angew. Chem., Int. Ed.*, 2004, **43**(26), 3400–3420.
54. B. L. Ryland and S. S. Stahl, Practical aerobic oxidations of alcohols and amines with homogeneous copper/TEMPO and related catalyst systems, *Angew. Chem., Int. Ed.*, 2014, **53**(34), 8824–8838.
55. A. N. Campbell and S. S. Stahl, Overcoming the "oxidant problem": strategies to use O2 as the oxidant in organometallic C–H oxidation reactions catalyzed by Pd (and Cu), *Acc. Chem. Res.*, 2012, **45**(6), 851–863.
56. B. H. Lipshutz and S. Ghorai, Transitioning organic synthesis from organic solvents to water. What's your E Factor? *Green Chem.*, 2014, **16**(8), 3660–3679.
57. B. H. Lipshutz, N. A. Isley, J. C. Fennewald and E. D. Slack, On the Way Towards Greener Transition-Metal-Catalyzed Processes as Quantified by E Factors, *Angew. Chem., Int. Ed.*, 2013, **52**(42), 10952–10958.
58. A. K. Agarwal, Biofuels (alcohols and biodiesel) applications as fuels for internal combustion engines, *Prog. Energy Combust. Sci.*, 2007, **33**(3), 233–271.
59. J. C. Liao, L. Mi, S. Pontrelli and S. Luo, Fuelling the future: microbial engineering for the production of sustainable biofuels, *Nat. Rev. Microbiol.*, 2016, **14**(5), 288–304.
60. X. Lu, B. Xiao, Z. Zhang, T. Gong, W. Su, J. Yi, Y. Fu and L. Liu, Practical carbon–carbon bond formation from olefins through nickel-catalyzed reductive olefin hydrocarbonation, *Nat. Commun.*, 2016, **7**, 11129.
61. K. C. Nicolaou, P. G. Bulger and D. Sarlah, Palladium-catalyzed cross-coupling reactions in total synthesis, *Angew. Chem., Int. Ed.*, 2005, **44**(29), 4442–4489.
62. R. L. Patman, J. F. Bower, I. S. Kim and M. J. Krische, Formation of CC Bonds via Catalytic Hydrogenation and Transfer Hydrogenation: Vinylation, Allylation, and Enolate Addition of Carbonyl Compounds and Imines, *Aldrichim. Acta*, 2008, **41**(4), 95.

63. J. F. Bower, I. S. Kim, R. L. Patman and M. J. Krische, Catalytic carbonyl addition through transfer hydrogenation: a departure from preformed organometallic reagents, *Angew. Chem., Int. Ed.*, 2009, **48**(1), 34–46.
64. M. Y. Ngai, J. R. Kong and M. J. Krische, Hydrogen-Mediated C–C Bond Formation: A Broad New Concept in Catalytic C–C Coupling 1, *J. Org. Chem.*, 2007, **72**(4), 1063–1072.
65. American Chemical Society. Green Chemistry: Solvents. http://www.acs.org/content/acs/en/greenchemistry/research-innovation/research-topics/solvents.html (accessed June 30, 2015).
66. U.S. Environmental Protection Agency. Presidential Green Chemistry Challenge Award Recipients by Technology Sector: Solvents. http://www2.epa.gov/green-chemistry/presidential-green-chemistry-challenge-award-recipients-technology#solvents (accessed June 30, 2015).
67. K. Sankula, S. Kota and S. Nissankarrao, Supercritical Fluid Technology: Green Chemistry for the 21st Century, *Pharma Innovation*, 2014, **3**(5), 19–24.
68. S. K. Ritter, Seeing the Green Side of Innovation, *Chem. Eng. News*, 2014, **92**(26), 24–28.
69. M. McCoy, Cleaning Product Makers Bask In New Solvents, *Chem. Eng. News*, 2015, **93**(3), 16–19.
70. S. K. Wisconsin Ritter, Depolymerized Lignin Makes A Good Solvent, *Chem. Eng. News*, 2012, **90**(24), 36.
71. Elevance Renewable Sciences. Our Product Lines, http://www.elevance.com/products/ (accessed July 7, 2015).
72. U.S. Environmental Protection Agency. Presidential Green Chemistry Challenge Awards: 2010 Small Business Award. http://www2.epa.gov/green-chemistry/2010-greener-synthetic-pathways-award (accessed July 7, 2015).
73. E. K. Wilson, A Cheaper-Than-Platinum Catalyst, *Chem. Eng. News*, 2012, **90**(34), 12.
74. W. C. Ellis, C. T. Tran, M. A. Denardo and A. Fischer, *et al.*, Communication: Design of More Powerful Iron-TAML Peroxidase Enzyme Mimics, *J. Am. Chem. Soc.*, 2009, **131**(50), 18052–18053.
75. B. S. Sekhon, Microwave-assisted pharmaceutical synthesis: an overview, *Int. J. Pharm. Tech. Res*, 2010, **2**(1), 827–833.
76. R. Wagner, Microwave-Assisted Synthesis in the Pharmaceutical Industry, *DDW Drug Discovery World*, 2006, **7**(3), 59.
77. I. Montes, D. Sanabria, M. García, J. Castro and J. Fajardo, A greener approach to aspirin synthesis using microwave irradiation, *J. Chem. Educ.*, 2006, **83**(4), 628.
78. F. Leonetti, C. Capaldi and A. Carotti, Microwave-assisted solid phase synthesis of Imatinib, a blockbuster anticancer drug, *Tetrahedron Lett.*, 2007, **48**(19), 3455–3458.
79. V. Polshettiwar and R. S. Varma, Microwave-assisted organic synthesis and transformations using benign reaction media, *Acc. Chem. Res.*, 2008, **41**(5), 629–639.

80. L. Pellegatti and J. Sedelmeier, Synthesis of Vildagliptin Utilizing Continuous Flow and Batch Technologies, *Org. Process Res. Dev.*, 2015, **19**(4), 551–554.
81. P. R. Murray, D. L. Browne, J. C. Pastre, C. Butters, D. Guthrie and S. V. Ley, Continuous flow-processing of organometallic reagents using an advanced peristaltic pumping system and the telescoped flow synthesis of (E/Z)-tamoxifen, *Org. Process Res. Dev.*, 2013, **17**(9), 1192–1208.

Subject Index